Vadose Zone Hydrology: Cutting Across Disciplines

sent by Scott Tyler

Groundwater Cleanup at Hanford — estimated at ~$80×10⁹ 80 billion dollars

EDITED BY DEBORAH SILVA

Newton-Raphson better than Picard iteration

UNIVERSITY OF CALIFORNIA, DAVIS
SEPTEMBER 1995

Tribute

James Wellington Biggar

Donald Rodney Nielsen

Vadose Zone Hydrology: Cutting Across Disciplines honors James Wellington Biggar and Donald Rodney Nielsen, two of the world's most distinguished soil scientists. For more than three decades, their collaborative research at the University of California, Davis, led to significant advances in solute transport and geostatistics. In the 1960s and 1970s, when the computer age was in its infancy and it was next-to-impossible to solve complex equations with multiple variables involving solute transport, Nielsen and Biggar pioneered the development of using mathematical models to interpret their experimental observations. Every student of solute transport has studied the landmark papers of Biggar and Nielsen that formalized miscible displacement theory. They demonstrated the coupled nature of mass flow and diffusion as well as the importance of chemical reactions during leaching. Their early studies established the groundwork for application of the theory of field-scale processes as both a research and management tool.

Nielsen and Biggar were among the first to describe the variability of soil properties in realistic statistical terms. Their systematic and thorough measurement of the spatial variability of soil-water contents, bulk densities, soil-water diffusivities, and solute dispersion coefficients, among other parameters, resulted in one of the most cited technical articles in the history of soil science. They showed for the first time that solute transport characteristics of field soils were largely log-normally distributed and developed a number of techniques for quantifying soil variability. Today, hundreds of scientists and engineers are using these techniques and developing even more advanced approaches for quantifying soil heterogeneity, with the aid of modern computer technology, based on the pioneering work of Biggar and Nielsen.

The seminal contributions of Nielsen and Biggar to advancing understanding of the physics of water movement and the chemistry of the interactions among solutes and soil were accomplished through teamwork. In 1986, Jim and Don were jointly awarded the Soil Science Research Award of the Soil Science Society of America (SSSA), the highest honor the society bestows for outstanding research contributions. Their award is unique because it represents the first and only time in the history of the SSSA that this award has been given to a team rather than an individual researcher. Parenthetically, as this tribute was being prepared, Jim suggested that "Biggar and Nielsen" appear in alternating rather than alphabetical order. It is a perfect example of the mutual respect, fairness, and spirit of equality that permeates their research partnership.

Don and Jim are more than top-notch researchers; they are outstanding educators. Together, they were the backbone of the teaching program in soil science at UC Davis, teaching courses in many subject areas at the undergraduate and graduate levels. Their expertise attracted graduate students and postdoctoral scholars from around the world. During their careers, Biggar and Nielsen supervised more than 60 master's and doctoral students and about 100 postdoctoral scholars. But the numbers are only part of the story. Don's and Jim's students loved them. Through their dedication and commitment to collaborative work and to serving as role models, Jim and Don motivated their students to strive for excellence and inspired them to achieve beyond their own expectations. Several of today's most prominent scientists in the field of solute transport were mentored by them. In the future, perhaps Don and Jim will have the special opportunity that Don Kirkham has at this meeting -- to present a poster at a conference honoring the career of a former student. (Don Nielsen received his Ph.D. in soil physics in 1958 under Kirkham's supervision at Iowa State.)

Scientists' understanding today of solute transport and geostatistics would not be nearly as far along were it not for the contributions of Jim Biggar and Don Nielsen. Their work has had a major impact on water management and environmental protection here and abroad. All of the participants of this international conference are proud to dedicate **Vadose Zone Hydrology: Cutting Across Disciplines** in honor of the unique research partnership of Don Nielsen and Jim Biggar.

Acknowledgments

It has been a year and a half since Marc and I (Jan) went to Wageningen, the Netherlands for a short meeting. With a stopover in Atlanta, the total flight time was at least 15 hours. Nevertheless, the trip was over before we knew it. The truth is, we decided to organize this conference during the flight and discussed its theme and contents for most of the flight's duration. It was then that we decided it should be a vadose zone conference and that its scope should be dedicated to Don Nielsen and Jim Biggar, whose research partnership has received much-deserved international acclaim for critical contributions to advancing the fields of solute transport and geostatistics for more than three decades.

We remember the surprised faces of passengers around us as we ceaselessly discussed options, raising one or both of our voices in preparing for this conference. I will never forget this experience! I was worthless for the next few days and purposely chose a seat away from Marc on the plane back to the U.S. I am not sure that he ever understood why. Still, the foundation for the conference was laid.

We were pleasantly surprised when Andrew Chang, Director of the Kearney Foundation of Soil Science, approved our conference proposal and fully sponsored the conference activities, including publication of the Proceedings. Thanks to the Kearney Foundation, we are able to compensate travel expenses for all of our invited speakers. We thank Rien van Genuchten, Andrew Chang, and Dennis Rolston for their input to the Proceedings. We are especially grateful to Walt Russell of the U.S. Salinity Laboratory for assembling all of the extended abstracts, and we also thank Deborah Silva for her contributions to the Proceedings. We acknowledge their gallant efforts to uphold the professional standards that we require, despite imminent publication deadlines, so that this document could be distributed on time during registration. We also give special thanks to the Division of Environmental Resource Sciences and Policy of the College of Agricultural and Environmental Sciences on the UC Davis campus for the financial assistance that has made it possible for some longtime students and professional colleagues to be present. We also thank Ken Tanji for his wise advice. At UC Davis, we are proud to have Conference Services, of which both Lina Caparas and Nesa Becker were instrumental in the preparation and organization. We also acknowledge the help of our administrative staff in Veihmeyer Hall. Thank you JoAnn, Janice, Tanya, and Catalina.

Finally, Don and Jim, we thank you for letting us organize the **Vadose Zone Hydrology: Cutting Across Disciplines** conference in your honor. We hope you both enjoy it. The conference will have been a success if it achieves an academic research caliber you can be proud of and if it meets the standard of excellence you established during your careers as outstanding educators and researchers at the University of California.

Jan Hopmans & Marc Parlange

Kearney Foundation of Soil Science International Conference
Vadose Zone Hydrology: Cutting Across Disciplines
September 6 - 8, 1995
Davis, California

September 6

5:00 - 8:00 p.m.   **Registration and Opening Reception**
Leach Hall Dormitory Courtyard

September 7

7:30 - 8:30 a.m.   **Registration and Poster Setup**
Freeborn Hall Lobby

8:30 - 8:45 a.m.   **Greetings and Welcome**
Freeborn Hall

J. W. Hopmans and M. B. Parlange
Hydrologic Science, Dept. of Land, Air and Water Resources, UC Davis

B. O. Schneeman
College of Agricultural and Environmental Sciences, UC Davis

8:45 - 10:15 a.m.   **Session 1a: Volume Averaging Across Scales**
W. A. Jury, Session Chair
Dept. of Soil & Environmental Sciences, UC Riverside

     8:45 a.m.   **Incorporation of Interfacial Areas in Models of Two-Phase Flow and Contaminant Transport**

M. A. Celia
Dept. of Civil Engineering, Princeton University

W. G. Gray
Dept. of Civil Engineering, University of Notre Dame

     9:30 a.m.   **Fundamentals of Transport Equation Formulation for Two-Phase Flow in Homogeneous and Heterogeneous Porous Media**

M. Quintard
L.E.P.T.-ENSAM (UA CNRS), France

S. Whitaker
Dept. Of Chemical Engineering, UC Davis

10:15 - 10:45 a.m.   **Coffee Break**
Freeborn Hall Lobby and Courtyard

10:45 - 12:15 p.m.   **Session 1b: Intermediate Scales**
W. A. Jury, Session Chair
Dept. of Soil & Environmental Sciences, UC Riverside

     10:45 a.m.   **Persistence and Interphase Mass Transfer of Organic Contaminants in the Unsaturated Zone: Experimental Observations and Mathematical Modeling**

L. M. Abriola, K. D. Penell, and W. J. Weber, Jr.
Dept. of Civil and Environmental Engineering, University of Michigan

11:30 a.m.   **Stochastic Differential Equations for Chemical Transport in Heterogeneous Porous Media**

Z. J. Kabala
Dept. of Civil Engineering, Duke University

G. Sposito
Dept. of Environmental Science, Policy and Management, UC Berkeley

12:15 - 1.15 p.m.   **Lunch**
Memorial Union

1:15 - 5:30 p.m.   **Session 2: Intermediate Scales**
W. A. Jury, Session Chair, Dept. of Soil & Environmental Sciences, UC Riverside

1:15 p.m.   **Diffusion-Linked Kinetics of Microbial Metabolism in the Vadose Zone**

W. R. Gardner
Professor Emeritus, UC Berkeley

R. F. Harris
Dept. Of Soil Science, University of Wisconsin, Madison

J. E. Waatsonn and Y. Liu
Dept. of Environmental Science, Policy and Management, UC Berkeley

2:00 p.m.   **Coupling Vapor Transport and Biodegradation of Volatile Organic Chemicals**

D. E. Rolston, K. M. Scow, and Y. El-Farhan
Hydrologic Science, Dept. of Land, Air and Water Resources, UC Davis

2:45 p.m.   **Refreshment Break**
Freeborn Hall Lobby and Courtyard

3:15 p.m.   **Transport of Reactive Solutes in the Subsurface: Coupling Reactions with Porous Media Heterogeneity**

M. L. Brusseau
Soil and Water Science Department, University of Arizona
Hydrology and Water Resources Department, University of Arizona

4:00 p.m.   **Infiltration and Water Movement in Soils**

J.-Y. Parlange
Dept. of Agricultural and Biological Engineering, Cornell University

R. Haverkamp
LTHE IMG, BP 53X, France

T. S. Steenhuis
Dept. of Agricultural and Biological Engineering, Cornell University

D. A. Barry
Center of Water Research, University of Western Australia

P. J. Culligan-Hensley
Ralph Parsons Laboratory, Massachusetts Institute of Technology

4:45 p.m.   **Passive and Active Microwave Remote Sensing of Surface Soil Moisture**

T. J. Jackson
USDA Agricultural Research Service Hydrology Laboratory, Beltsville, MD

E. T. Engman
NASA, Goddard Space Flight Center, Hydrological Sciences Branch, Greenbelt, MD

T. J. Schmugge
USDA Agricultural Research Service Hydrology Laboratory, Beltsville, MD

7:00 - 10:00 p.m.   **California Barbecue** at Putah Creek Lodge

September 8

7:30 - 8:00 a.m.   **Poster Setup**

8:00 - 12:15 p.m.   **Session 3: Field Scales**
D. McLaughlin, Session Chair
Ralph Parsons Laboratory, Massachusetts Institute of Technology

8:00 a.m.   **Water and Solute Transport in Arid Vadose Zones: Innovations in Measurement and Analysis**
S. W. Tyler
Desert Research Institute, Reno, Nevada
B. R. Scanlon
Texas Bureau of Economic Geology, Austin
G. W. Gee
Batelle Pacific Northwest Laboratories, Richland, Washington
G. B. Allison
CSIRO Division of Water Resources, Australia

8:45 a.m.   **Recent Advances in Vadose Zone Flow and Transport Modeling**
E. A. Sudicky
Waterloo Centre for Groundwater Research, University of Waterloo
M. Th. van Genuchten
U.S. Salinity Laboratory, USDA Agricultural Research Service, Riverside, CA

9:30 a.m.   **Water Flow in Desert Soils near Buried Waste Repositories**
P. J. Wierenga, A. W. Warrick, and L. Pan
Dept. of Soil and Water Sciences, University of Arizona

10:15 a.m.   **Coffee Break** in Freeborn Hall Lobby and Courtyard

10:45 a.m.   **Site-Specific Management of Flow and Transport in Heterogeneous and Structured Soils**
D. J. Mulla
Dept. of Soil, Water, and Climate, University of Minnesota
O. Wendroth and M. Joschka
ZALF, Muncheberg, Germany

11:30 a.m.   **Field-Scale Modeling and GIS Technology as Tools in Evaluating Solute Fluxes in Agro-Ecological Systems**
R. J. Wagenet
Dept. of Soil, Crop and Atmospheric Sciences, Cornell University
J. Bouma
Dept. of Soil Science and Geology, Wageningen Agricultural University

12:15 - 1:45 p.m.   **Lunch**

1:45 - 4:30 p.m.   **Poster Session**
Freeborn Hall

4:00 — 5:00
4:30 - 6:00 p.m.   **Present Directions and the Future Research of Vadose Zone Hydrology — Presentation and Participative Discussion**
W. A. Jury
Dept. of Soil & Environmental Sciences, UC Riverside
D. McLaughlin
Ralph Parsons Laboratory, Massachusetts Institute of Technology

6:30 - 10.00 p.m.   **Closing Reception and Dinner**
Freeborn Hall Courtyard

Extended Abstracts

[Editor's Note: The number assigned to each Extended Abstract corresponds to the number on display on the poster board. An alphabetical index of all presenters' names and their corresponding poster numbers is on p. 171-172.]

Appendix

Scaling the Kinetics of Soil Aggregate Breakdown

J.D. Albertson[1], E. Zanini[2], E. Bonifacio[2], and D.R. Nielsen[1]. [1]*Hydrologic Science, University of California, Davis, CA 95616 USA.* [2]*Dipartimento di Valorizzazione e Protezione delle Risorse Agroforestali, Università di Torino, I-10100 Torino, Italy.*

Introduction. Soil erosion is of fundamental importance to agriculture as well as the management of down-slope resources. However, to develop a theoretical, predictive model of erosion would require the inclusion of a wide array of independent variables, ranging from landscape geometry to soil mechanical properties to land-use history characterization. In the present effort we seek to collapse this information into a single scale factor. Toward this objective we analyze soil aggregate breakdown time series in the context of the extended scaling framework of Simmons, Nielsen, and Biggar (1). The data collection and treatment procedures are described elsewhere in this volume (2). As described in (2), the aggregate breakdown time series are represented by

$$Y'(t) = Y(t) - a = b[1 - \exp(-t/c)] \tag{1}$$

where $Y(t)$ is the cumulative aggregate mass loss, t is duration of wet sieving, and a, b, and c are parameters for the specific sample .

Materials and Methods. In the present work we analyze the $Y'(t)$ series which represent the measured data minus their zero time intercept, which is due to explosion of the aggregate upon initial wetting. The unscaled $Y'(t)$ time series for 30 samples are shown in Figure 1. The great variability in the series is due to the different samples representing different locations with varying slope, soil type, parent material, and land use. For practical purposes we seek a single scale factor for each sample to represent the net effect of these many independent variables. The scaling framework is based on the work of Miller and Miller (3) and is extended and described in detail by (1). Here we examine its extension to soil aggregate breakdown. We seek a scale mean curve $Y.'(t)$, such that each time series $Y_i'(t)$ can be represented as a linear function of $Y.'(t)$, viz.

$$\lambda_i Y_i'(t) = Y.'(t) \tag{2}$$

The scale mean time series $Y.'(t)$ are derived from

$$\frac{1}{Y.'(t)} = \sum_{i=1}^{N} \frac{1}{Y_i'(t)} \tag{3}$$

where N is the number of time series. The scale factors are obtained by minimizing the squared misfit

$$Misfit = \sum_t \sum_{i=1}^{N} [\lambda_i Y_i'(t) - Y.'(t)]^2 \tag{4}$$

subject to the constraint

$$\frac{1}{N}\sum_{i=1}^{N} \lambda_i = 1.0 \tag{5}$$

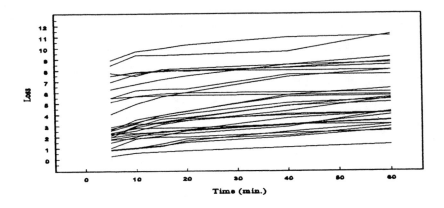

Fig. 1. Unscaled Aggregate Breakdown Time Series

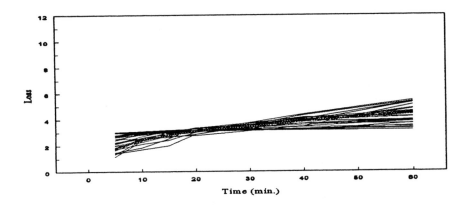

Fig. 2. Scaled Aggregate Breakdown Time Series

Results and Discussion. The scale mean curve was computed from the 30 time series using [3] and the scale factors λ_i derived from a forward Newton iteration using [4] and [5]. The scaled time series $\lambda_i \Upsilon'(t)$ are shown in Figure 2.

Conclusion. Much of the variability between samples has been effectively captured by a single scaling parameter, as shown by the collapse of the time series between Figures 1 and 2. Future work will seek a means of estimating a soil's scaling parameter from simple site specific indices representing physical, chemical, and biological factors.

Literature Cited.
(1) Simmons, C.S., D.R. Nielsen, and J.W. Biggar, 1979, Scaling of field-measured soil-water properties: I. Methodology, and II. Hydraulic conductivity and flux, Hilgardia, 47, 77-173.
(2) Zanini, E., E. Bonifacio, and D.R. Nielsen, 1995, Comparison of aggregate breakdown in pasture topsoils, Elsewhere in this volume.
(3) Miller, EE., and R.D. Miller, 1956, Physical theory for capillary flow phenomena, J. Appl., Phys., 27, 324-332.

Simulation of Soil Evaporation Using an Energy Balance Model Approach

G. Alvenäs. *Department of Soil Sciences, Swedish University of Agricultural Sciences, P.O. Box 7014, S-750 07 Uppsala, Sweden.*

Introduction. Soil evaporation is one of the main processes in the complex exchange of heat and water at the soil surface. It is regulated by atmospheric conditions, energy supply, surface moisture and moisture transport in the soil. Therefore, solvation of the equation for soil evaporation requires accurate estimates of temperature and moisture conditions at the soil surface. Several simulation models using different formulas for the moisture availability at the surface have been presented (e.g. 1 and 2). Commonly, surface moisture is substituted by the relative humidity of the air in equilibrium with the water in the soil pore. Based on thermodynamic relationships Philip (4) expressed the surface humidity as a function of soil water potential (Y_s) and temperature of the surface. One question to be answered is how to determine Y_s from soil water potential in the uppermost layer of the soil (Y).

This poster presents a model approach where Y is multiplied by a semi-empirical correction term, based on surface water balance, that accounts for the steep moisture gradient in the uppermost part of the soil. Simulated soil water content and temperature was compared with measurements and the significance of the correction term was tested.

Materials and Methods. A small-scale field experiment was set up, comprising two clay plots and two sand plots, each of them 2 x 2 m and 0.3 m deep. One sand plot and one clay plot were irrigated during dry periods while the other plots were dry. The plots were kept bare and were surrounded by grass that was kept short. A number of different measurements were carried out. Besides the standard meteorological measurements at the site, soil temperature and soil water content and tension were registered continuously in profiles in each plot. An IR thermometer provided data on surface temperature. Surface moisture was determined by means of a gas analyser.

One approach of the SOIL model (3) solves the heat flow equations at the soil surface. In the calculations of latent heat, surface vapour pressure (e_{ss}) is calculated from surface temperature and water tension in the uppermost part of the soil (Y):

$$e_{ss}=e_s(T_{ss})\exp(\frac{-\psi e_{corr}}{RT})$$

[1]

where $e_s(T_{ss})$ is the saturated vapour pressure at the temperature of the surface, T the temperature and R the gas constant. e_{corr} is an empirical correction term accounting for steep gradients in moisture between the uppermost soil layer and the surface. The correction term is a function of the the water balance of the surface layer (D_{surf}), limited to vary between -2 and 1 mm of water and an epirical factor (Y_{eg}):

$$e_{corr}=10^{(-\Delta_{surf}\psi_{eg})}$$

[2]

Results and Discussion. A test of the the significance of the correction factor, Y_{eg}, was made for a sand soil during a dry period including two irrigation events (fig. 1). Comparisons were made between a simulation without any correction ($Y_{eg}=0$) and a simulation where the moisture availability in the surface layer could decrease with two powers of magnitude during drying, and increase with one power of magnitude during wetting ($Y_{eg}=1$). During dry periods, the correction influenced both surface temperature and evaporation significantly (figs 2 and 3). Because of dryer surface conditions using the correction term, evaporation decreased during daytime. The correction also caused an decrease in night-time condensation. Surface temperature predicted using the correction term, agreed better with measured values as compared with the prediction where no correction term was used. Soil water content differed slightly between the two predictions but only in the uppermost soil layers. Further tests of the correction term will be made using data from different moisture conditions for both clay and sand.

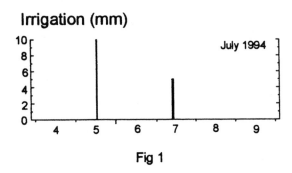

Fig 1

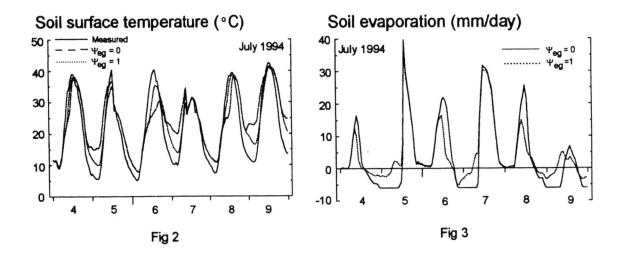

Fig 2

Fig 3

Conclusion. Including a correction term in the vapour pressure expression as described by the law of thermodynamics, is of great significance for the prediction of near-surface conditions. Substantial improvements of the predictions of surface temperature can be made. Both evaporation and condensation are affected by the correction. Futher analyses for different soils and different moisture conditions will be made to investigate the importance of the soil properties for the surface vapour pressure and evaporation.

Literature Cited.
(1) Camillo, P.J., R.J. Guerney and T.J. Schmugge. 1983. A soil and atmospheric boundary layer model for evapotranspiration and soil moisture studies. Water Resourc. Res. 19: 371-380.
(2) Chung, S.-O., and R. Horton. 1987. Soil and heat water flow with a partial surface mulch. Water. Resourc. Res. 12: 2175-2186.
(3) Jansson, P.-E. 1991. Simulation model for water and heat conditions. Description of the SOIL model. Swedish Univ. of Agric. Sciences, Dept. of Soil Sciences, Div. of Agric. Hydrotechnics, report 165.
(4) Philip, J.R. 1957. Evaporation, and moisture and heat fields in the soil. J. Meteor. 14:354-36.

Monitoring of Soil Water and Solute Distribution in an Almond Orchard

L. Andreu, J.W. Hopmans, T. Frueh, L.J. Schwankl, A. Tuli, S. Essert and J. Macintyre. *Hydrologic Science, 113 Veihmeyer Hall, University of California Davis, CA 95616.*

Introduction. Scheduling and management of high frequency irrigation systems can be improved by soil water and solute monitoring, thereby reducing water and fertilizer losses and avoiding salinity problems. The Measurement of water and solute distribution in orchards with high frequency irrigation systems requires intensive instrumentation because of the three dimensional pattern of water and salts (1,2). In earlier studies, water monitoring includes soil coring, tensiometers, gypsum blocks and neutron probes, while solute concentrations are measured by suction sampling or soil coring. Recently, TDR has shown to be a localized and accurate method for in-situ measurement of soil water and salinity. In this paper we use several methods for monitoring different irrigation systems in an almond orchard. We propose a combined tension-solution probe which measures soil water pressure and extract solution simultaneously. A new TDR technique (3) permitting vertical soil moisture measurement was calibrated in the laboratory for future field use. Our objective is to achieve a good understanding of water and solute transport under drip irrigation and to develop practical guidelines for proper management of these systems.

Materials and Methods. A field experiment was conducted at Nickels Ranch, 40 miles north of Davis, in a five-year-old almond orchard with trees planted in a 4.8 m X 6.6 m pattern . The experiment included four varieties (Butte, Carmel, Nonpareil and Monterrey) and four irrigation systems (surface drip, microsprinklers, sub-surface drip single line and surface drip double line). Irrigation amounts were estimated using potential evapotranspiration and a crop factor based on percent canopy cover and time of the year. In the surface drip system, thirty-one access tube (1.35 m depth) were installed on one side of a single tree (Fig 1). Soil water content was measured at depths of 0.15, 0.30, 0.45, 0.60, 0.75, 0.90, 1.05 and 1.20 m by the neutron probe . Initial soil sampling for texture and salinity was carried out for the access tubes locations. Forty-six porous combined solution sampler-tensiometer probes were built and installed at two depths (1.05 and 1.35 cm). In single line sub-surface drip, twenty five access tube, with the same characteristic as the surface treatment, were installed around one tree. In the microsprinkler treatment, a tree was instrumented with gypsum blocks at a regular grid at two depths , 0.20 and 0.90 m, to measure changes in soil water contents. Measurements were taken every week in 1994. Multi-steps outflow method and in situ experiments were used for soil characterization. Root excavations were carried out at the end of the growing season to observe the root tree characteristics. A 1.5 m long soil column from the experimental farm was used in the laboratory for calibration of the TDR probes.

Results and Discussion. Analysis of water patterns show variability due principally to the relative distance to the emitters and the distribution of water uptake by the trees. Hydraulic conductivity and gradients measurements at the bottom of the root zone allowed estimation of drainage fluxes. We noticed the presence of leaching areas where the emitters apply water at a rate higher than the tree takes up and in soil regions with no roots. We conclude that emitter position and discharge is important for efficient irrigation management. Otherwise, loss of water and nutrients can be significant. Root surveys showed that roots reached 0.90 m depth. In the drip treatment, roots are concentrated along the dripper line and close to the tree. In the microsprinkler treatment , roots are more spread laterally. Salinity measurements obtained with soil solution samplers showed a decrease in salinity at 1.05 and 1.35 m depth below the emitters during the irrigation season. The combined tensiometers/solution samplers were useful for the simultaneously monitoring of soil water tension and salinity.

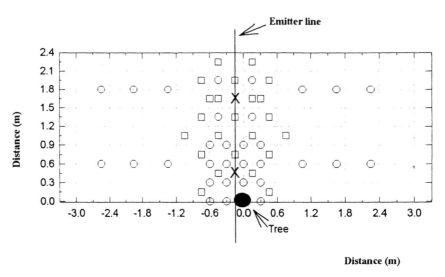

Fig 1. Set up of a instrumented tree

Literature Cited.
(1) Rolston, D.E., J.W. Biggar, and H.I. Nightingale. 1991. Temporal persistence of spatial soil-water patterns under trickle irrigation. Irrig. Sci. 12:181-186.
(2) Nightingale, H.I., G.J. Hoffman, D.E. Rolston and J.W. Biggar. 1991. Trickle irrigation rates and soil salinity distribution in an almond orchard. Agric. Water Manage. 19:271-283.
(3) Hook, W.R., N.J. Livingston, Z.J. Sun, and P.B. Hook. 1992. Remote diode shorting improves measurement of soil water by time domain reflectometry. Soil Sci. Soc. Am. J. 56:1384-1391.

An Inexpensive and Easy-to-use System to Automate the Collection and Analysis of Soil Physical Infiltration and Outflow Measurements

Mark D. Ankeny, Mark A. Prieksat, and Mark E. Burkhard. *Daniel B. Stephens and Assoc., Inc., 6020 Academy Road, Albuquerque, New Mexico 87109, USA.*

Introduction. An integrated system, capable of making testing decisions, recording data, and extracting parameters automatically would be useful in many soil physics projects. We have integrated infiltration and outflow devices with inexpensive datalogging capabilities, and automated analysis in Excel to facilitate experimentation. We have developed a simple, inexpensive datalogger that operates in DOS based BASIC-52 programming language. The datalogger can operate electronic valves and switches, as well as communicate with peripheral devices such as pressure transducers, temperature probes, electronic balances, and almost any other device. The logger was used to automate a one-step outflow cell. In this scheme the logger communicates with a balance, operates a solenoid valve, and analyzes voltage inputs from pressure transducers then converts them to actual pressure readings. The logger has the ability, through flexible programming language and direct communication with the balance, to scan input from the balance and to only record data when there has been some predetermined change in the balance reading. This feature allows the acquisition of information as opposed to merely collecting data.

Materials and Methods.
Datalogger Design: The basic design criteria for the datalogger were (1) logical programming language, (2) user specified data acquisition and peripheral device control, and (3) low cost. The datalogger has an eight channel 12 bit analog-to-digital converter. Each channel is equipped with a differential amplifier circuit with resistor programmable gain. Each channel is also equipped with both 5 volts DC and ground. Thus, each channel can be tailored to fit a user-specified function. The unit is equipped with one bidirectional serial port and one serial output port. The datalogger communicates with the computer during the programming stage via the bidirectional RS-232 port. The other port is used for sending program listings to a printer. However, during operation, the ports can have other designated uses such as communicating with a balance or other RS-232 capable devices. The unit can store up to 32k bytes of data. All data storage is written to battery backed RAM so that the data is not lost if power is interrupted. Data is stored and retrieved under program control. Therefore, virtually any required data output format may be implemented. It is also possible to do data analysis and output only required values, such as hydraulic conductivity (K). The unit is programmed in INTEL MCS BASIC-52 language from a terminal of computer connected via the RS-232 port. Programs can be up to 28k bytes in size. The unit can be used with any 8 to 20 volt DC power supply, or a set of two or three D-cell batteries.

Operation of a One-step Outflow Cell: Figure 1 shows a schematic diagram of the one-step outflow cell. The datalogger was programmed to actuate the small solenoid valve, communicate with the balance, and monitor the pressure transducer under preset conditions and restraints. Starting with a nearly saturated soil core, the core is allowed to free drain, then the system is pressurized. Outflow is monitored by weight changes in the system as communicated by a balance, and surface potential is monitored via a tensiometer connected to a pressure transducer. Flow is controlled by opening or closing the solenoid valve. The datalogger was programmed so that the test was initiated by depressing the tare bar on the balance. The user is prompted to press the tare bar. The first press of the tare bar opens the solenoid valve and allows water flow from the system. When free drainage has ceased, the tare bare is pressed again and the valve closes. After pressure is applied to the system, the tare bar is pressed again reopening the valve which allows outflow from the system. Data monitoring is continuous throughout the experiment, but data is stored on a continuous time basis only for the first 20 seconds of the run. After that time, data is stored for user specified changes in weight as communicated by the balance. Data output contains (1) elapsed time in seconds, (2) change in weight, and (3) soil-surface potential.

Results and Discussion. Figure 2 shows outflow data collected for a Nicollette soil. The data is smooth and clean, showing uniform information content as opposed to time-based data collection. Rapid outflow rates generally taper off after the first hour of flow. Because of space, the surface potential vs time plot was eliminated, but it too showed smooth, consistent data. Together the outflow data and the surface water potential can be used to determine soil hydraulic properties (1). The entire soil sample is represented by the

the boundary observations and provides a best possible determination of the soil hydraulic properties. The datalogger was effective at operating the system with minimal human input. Data acquisition and analysis was effective, and clean data functions were obtained without externally manipulating the data. This datalogger allows flexibility and ease of use, at a cost substantially below any currently available datalogger. The estimated cost of the datalogger is around 400 dollars.

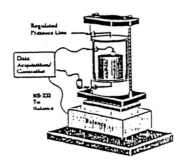

Fig. 1. One-Step Outflow Cell schematic

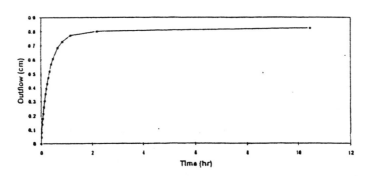

Fig. 2. Outflow vs time for a Nicollette soil

Literature Cited.
(1)   Ankeny, M.D., T.C. Kaspar, and R. Horton. 1987. Design for an automated tension infiltrometer. Soil Sci. Soc. Am. J. 52:893-896.
(2)   Kool, J.B., and J.C. Parker. 1987. Estimating soil hydraulic properties from one-step outflow experiments by parameter estimation. I. Theory and numerical studies. Soil Sci. Soc.Am. J. 49:1348-1354.

Water Uptake by Swelling Aggregates

E. Braudeau[1]. [1] Institut Français de Recherche Scientifique pour le Développement en Coopération (ORSTOM), B.P. 1386, Dakar, Senegal.

Introduction. In a recent work (1), experimental results on water uptake by aggregates are presented and interpreted by the Green-Ampt (2) approach. However, these results are obtained on a rigid soil matrix of oven stabilized clay aggregates. Since a great variety of natural soils swell (shrink) upon wetting (drying), we present here experimental results on water uptake by swelling aggregates and interpret them on the light of a double porosity material concept (3).

Materials and Methods. Several soil clods are withdrawn from the B horizon (40-60 cm depth) of an oxysol. Half of the clods served to obtain undisturbed soil cores 5.6 cm diameter, 3 cm height , the other half was fractioned through a 2 mm sieve. Samples are wetted in a beaker as shown schematically in fig.1a. Measurements are made with a time step of 100 seconds during the first 2 hours of swelling and of 300 seconds after; and are terminated when the height of the sample remains constant for at least 2 hours. The saturated samples are then dried in order to determine their shrinkage curve (3). A typical one is shown on fig. 2 with the definitions of the characteristic points.

Results and Discussion. Figure 1b shows experimental data of swelling (curve sw) and subsequent shrinkage (curve sh) of sample T2 in Table1. To model the swelling of aggregates, we assume that they are subject to a "swelling pressure" expressed as:

$$P_s = E/(\theta_\mu - v_{p\mu^\circ}) = E/\theta_{\mu no} \qquad [1]$$

where E is "the potential energy of the solid phase resulting from the surface charge of the soil in joules/kg" (4) and $\theta_{\mu no}$ is the water content soaked up by the micro-organized aggregates, which corresponds to the available micro pore volume, and:

$$d\theta_{\mu no}/dt = -K(P_s - P_s^\circ) \quad \text{with} \quad P_s^\circ = E/\theta_{\mu no}^{max} \qquad [2]$$

Considering the assumptions of the shrinkage model (3):

$$dv/dt = K_b d\theta_{\mu no}/dt \qquad [3]$$

integration of [3] yields:

$$\ln((v^{max} - v)/\gamma) + 1 - (v^{max} - v)/\gamma = A.t \qquad [4]$$

Fitting [5] on experimental data gives A and γ shown on fig.3. Considering [3], eq. [4] can be expressed as:

$$(1-S) \cdot \exp(S) = \exp(At) \qquad [5]$$

where S is the saturation degree of aggregates: $S = \theta_{\mu no}/\theta_{\mu no}^{max} = 1 - (v^{max} - v)/\gamma$.

Table 1 reveals that water absorbed by swelling is far from negligible since $\theta_{\mu no}^{max} \approx (\theta_{MS} - \theta_{AE})/2$ and approximatly half the residual water content $\theta_{SL}/2$. For swelling soils, the kinetics of absorption is

sample	clay	θ^{SL}	θ^{AE}	θ^{LM}	θ^{MS}	θ^{sat}	v^{SL}	v^{MS}	v^{sat}	vp_μ^{MS}	vp_μ^{SL}	$\theta_{\mu no}^{max}$	K_b
	%						cm³g⁻¹						
T1	46.1	0.117	0.142	0.153	0.194	0.334	0.677	0.704	0.705	0.170	0.132	0.038	0.672
R1T1		0.120	0.148	0.162	0.203	0.401	0.752	0.770	0.773	0.179	0.136	0.043	0.404
R2T1		0.112	0.136	0.157	0.190	0.398	0.752	0.768	0.770	0.171	0.126	0.045	0.356
T7	22.9	0.053	0.086	0.099	0.157	0.280	0.629	0.650	0.656	0.123	0.072	0.051	0.384
R1T7		0.048	0.081	0.100	0.135	0.291	0.648	0.662	0.667	0.115	0.067	0.047	0.319
R2T7		0.064	0.071	0.094	0.135	0.285	0.648	0.659	0.661	0.111	0.068	0.043	0.268
T9	10.4	0.016	0.036	0.060	0.101	0.267	0.637	0.645	0.646	0.077	0.028	0.050	0.160
R1T9		0.024	0.044	0.061	0.087	0.229	0.600	0.605	0.608	0.072	0.036	0.036	0.127
R2T9		0.036	0.044	0.068	0.099	0.223	0.600	0.601	0.602	0.081	0.041	0.040	0.049

Table 1: Characteristic points of the shrinkage of the undisturbed samples (T1,T7 and T9) and the corresponding fractioned sample for two cycles of swelling and shrinkage.

much more slower than that of rigid soils (1). Finally, it seems that the water potential in swelling aggregates is due to osmotic forces (proportional to 1/v) rather than capillary forces (proportional to 1/r).

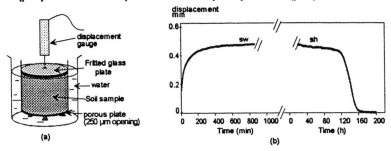

Figure 1: Sketch of the experimental apparatus (a) and experimental data (b) of swelling (curve sw) and shrinkage (curve sh) of sample T1.

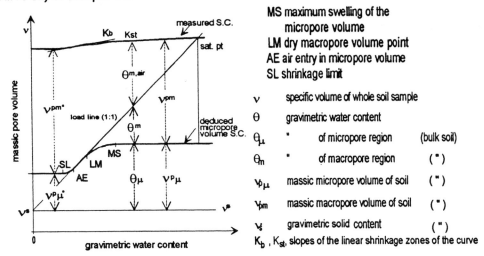

MS maximum swelling of the micropore volume
LM dry macropore volume point
AE air entry in micropore volume
SL shrinkage limit

v	specific volume of whole soil sample	
θ	gravimetric water content	
θ_μ	" of micropore region	(bulk soil)
θ_m	" of macropore region	(")
vp_μ	massic micropore volume of soil	(")
vp_m	massic macropore volume of soil	(")
v_s	gravimetric solid content	(")

K_b , K_{st}, slopes of the linear shrinkage zones of the curve

Figure 2: Typical shrinkage curve and definitions of the various characteritic points.

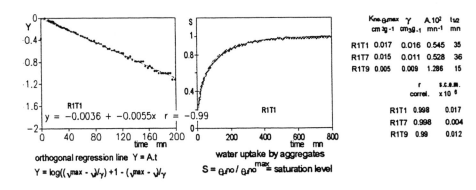

	Kna	$\theta_n$max	γ	A.10^2	t$_{1/2}$
	cm^{3}g^{-1}	cm^{3}g^{-1}		mn^{-1}	mn
R1T1	0.017	0.016	0.545		35
R1T7	0.015	0.011	0.528		36
R1T9	0.005	0.009	1.286		15

	r correl.	s.c.o.m. x 10^6
R1T1	0.998	0.017
R1T7	0.998	0.004
R1T9	0.99	0.012

orthogonal regression line Y = A.t

$Y = \log((v^{max} - v)/\gamma) + 1 - (v^{max} - v)/\gamma$

water uptake by aggregates

$S = \theta_{no} / \theta_{no}^{max}$ = saturation level

Figure 3: Results of the fit of equation [5] on experimental data.

Litterature cited.

(1) Youngs E.G., P.B. Leeds-Harrisson, and R.S. Garnett. 1994. Water uptake by aggregates. Europ. J. Soil Sci., 45 : 127-134.

(2) Green W.H., and G.A. Ampt. 1911. Studies in soil physics. I. The flow of air and water through soils. J. Agric. Sci., 4 : 1-24.

(3) Braudeau E. 1988. Equation généralisée des courbes de retrait d'échantillons de sol structurés. C. R. Acad. Sci. Paris, 207, série II, pp 1731-1734.

(4) Voronin A.D. 1983. An energetic approach to the quantitative evaluation of soil structure. Soviet Soil Science, 103-109.

Modelling Shrinkage of Unconfined Soil Cores

E. Braudeau[1], and **J. Toumâ**[?] *Institut Français de Recherche Scientifique pour le Développement en Coopération (ORSTOM), B.P. 1386, Dakar, Senegal, [2]ORSTOM, B.P. 5045, 34032 Montpellier Cedex 5, France.*

Introduction. The shrinkage curve (SC) relates the apparent specific volume of a soil core to its gravimetric water content. It is aknowledged to be an important characteristic of the soil, synthesizing informations on the soil pore space deformation as a function of its water content (1). In this paper we present a method to determine the SC and a new model to describe it.

Materials and methods. Saturated cores (5.6 cm diameter, 3 cm height) are placed in an isothermal chambre (30 ± 0.5 °C) each on the pan of a digital balance and the tip of a displacement transducer on each. Measurements are made with a time step of 5 minutes and are terminated when the core's diameter remains constant for at least 10 hours. Afterwards, the solid mass, M_s (M), apparent specific volume, v_t ($L^3.M^{-1}$), to which corresponds diameter D_t (L), and specific volume of solids v_s ($L^3.M^{-1}$) are determined independently. Assuming isotropy, the SC is determined by $v = v_t[D(\theta)/D_t]^3$, θ ($M.M^1$) being the gravimetric water content. Figure 1a shows the SC of sample S1 (Table 1), with θ_{sat} the saturated water content defined as the intersection of the SC with the load line (2) (slope $1/\rho_w$ and intercept at v_s, ρ_w being the density of water). Two domains are distinguished (3): the micro-organized clayey aggregates, (subscript μ), and their assembly with the other soil constituents (subscript m). Assuming that dv is a linear combination of $d\theta_m$ and dv_μ according to:

$$dv = \rho_w.K_b.dv_\mu + K_{st}.d\theta_m \qquad [1]$$

K_b and K_{st} being the slopes of the basic (4) and the linear part of stuctural shrinkage stages respectively, and that the point AE on the SC defines the "air entry" point in the micro-organized aggregates; it can be shown (5) that points SL, LM and MS correspond respectively to: the shrinkage limit, the limit of contribution of macroporosity to shrinkage, and the maximum swelling of the micro-organized aggregates. Equation [1] allows also to deduce the variation of the micro pore volume, $v_{\mu p}$:

$$v_{\mu p} - v_{\mu p,AE} = (v - v_{AE})/(\rho_w.K_b) \qquad \text{for } \theta < \theta_{AE} \qquad [2]$$

$$v_{\mu p} - v_{\mu p,AE} = \{v - v_{AE} - K_{st}(\theta - \theta_{AE})\}/\{\rho_w.(K_b - K_{st})\} \qquad \text{for } \theta > \theta_{AE} \qquad [3]$$

with $v_{\mu p,AE} = \theta_{AE}/\rho_w$, shown on fig. 1b for sample S1, with the origin of $_\mu v_p$ taken at v_s. The three linear parts distinguished on fig 1a are simply modelled by: $dv/d\theta = C$ with $C = 0$ for $\theta < \theta_{SL}$, $C = K_b$ for $\theta_{AE} < \theta < \theta_{LM}$, and $C = K_{st}$ for $\theta > \theta_{MS}$. For $\theta_{SL} < \theta < \theta_{AE}$ a very good fit on the observed data is obtained by :

$$dv/d\theta = K_b.[exp(\Theta_r)-1]/[exp(1)-1] \qquad [4]$$

where $\Theta_r = (\theta - \theta_{SL})/(\theta_{AE} - \theta_{SL})$. It is integrated readily and yields :

$$v = v_{SL} + (\theta_{AE} - \theta_{SL}).\{K_b[exp(\Theta_r) - \Theta_r - 1]\}/[exp(1)-1] \qquad [5]$$

For $\theta_{LM} < \theta < \theta_{MS}$, the following expression is fitted on the observed data:

$$dv/d\theta = K_b.[exp(\Theta_s)-1]/[exp(1)-1] + K_{st}.\{1-[exp(\Theta_s)-1]/[exp(1)-1]\} \qquad [6]$$

where $\Theta_s = (\theta_{MS} - \theta)/(\theta_{MS} - \theta_{LM})$. Integration of [6] gives :

$$v = v_{MS} - (\theta_{MS} - \theta_{LM}).\{K_b.[exp(\Theta_s)-\Theta_s-1] + K_{st}.[\Theta_s.exp(1)-exp(\Theta_s)+1]\}/[exp(1)-1] \qquad [7]$$

Figure 2 shows results of the model applied on soils listed in Table 1. Compared to observations, the fit is very satisfactory.

11

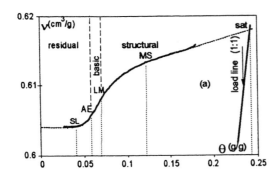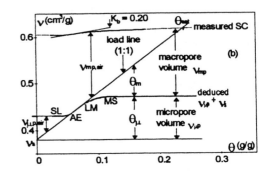

Figure1: Measured SC for sample S1 in Table 1 (a), and deduced SC of the micro-organized aggregates (b), based on assumptions in the text.

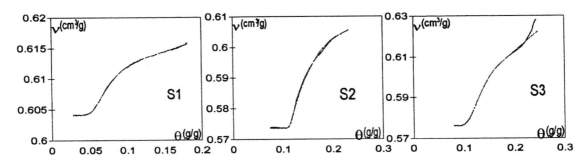

Figure 2: Results of the present model applied to samples listed in Table 1.

sample	Soil	% Clay	SL	AE	LM	MS	sat	K_{st}, K_b
S1	Oxysol	22.9	0.040	0.058	0.069	0.121	0.240	0.039
			0.604	0.606	0.608	0.613	0.616	0.197
S2	Ultisol	385	0.106	0.116	0.120	0.211	0.220	0.081
			0.574	0.576	0.578	0.604	0.604	0.554
S3	Sulfaquent	418	0.091	0.116	0.134	0.177	0.245	0.210
			0.576	0.856	0.592	0.607	0.622	0.570

Table1: Coordinates of the characteristic points θ (upper number) and v and slopes of the structural (upper number) and basic shrinkage stages for various soil types.

Litterature cited

(1) Mc Garry D., 1988: Quantification of the effects of zero and mechanical tillage on a vertisol by using shrinkage curve indices. Austr. J. Soil Res., 26:537-542.
(2) Sposito G., 1972: Thermodynamics of swelling clay-water systems. Soil Sci., 114:242-249.
(3) Brewer R., 1964: Fabric and Mineral Analysis of Soils. John Wiley and Sons, New York.
(4) Mitchell A.R., 1992: Shrinkage terminology: Escape from "Normalcy". Soil Sci. Soc. Am. J., 56:993-994.
(5) Braudeau E., 1988: Equation généralisée des courbes de retrait d'échantillons de sol structurés C. R. Acad. Sci. Paris, 207, série II, pp 1731-1734.

Calculation and Prediction of Scalar Roughness Lengths above a Drying Soil

A.T. Cahill, M.B. Parlange, J.D. Albertson. *Department of Hydrologic Science, University of California, Davis, CA 95616.*

Introduction. Monin-Obukhov similarity theory is often used to predict the fluxes of mass and energy from soils to the atmosphere. A necessary value for the use of the Monin-Obukhov equations is the roughness length for the scalar in question. However, since the scalar roughness length arises as an integration constant, it cannot be directly measured. It has been theorized that for bare soils scalar roughness lengths are functions of the Reynolds roughness number, z_{0+} [1,2] of the form

$$\ln\left(\frac{z_{0h}}{z_0}\right) = a + b z_{0+}^c \qquad [1]$$

where a, b and c are constants determined by the geometry of the surface. The use of various theoretical models and laboratory-based calibrations of this relationship has underestimated heat (H) and moisture fluxes (E) from bare soils when applied in the field. [3] An experiment was undertaken to develop a field-based relationship between either the scalar roughness length for heat, z_{0h}, and for water vapor, z_{0v}, and z_{0+}. The resulting measurements are analyzed by weighted regression analysis for a non-linear best-fit.

Materials and Methods. The soil flux, temperature and moisture data used in the analysis was taken in two experiments during the summer of 1994. The first experiment took place at the University of California's Campbell Research Tract in Davis, California, while the second was conducted on the dry bed of Owens Lake in Owens Valley, California. A sprinkler irrigation system was used to irrigate the field three times during the course of the experiment. The soil surface was saturated for at least a day after the irrigation. There was no natural precipitation during the experiment. The soil of the Campbell Tract is Yolo clay loam The Owens Lake bed is very arid, and a salt crust on the soil surface prevents much evaporation from the groundwater. [4] Quantities measured included H and E, net radiation, air temperature and humidity, soil temperature, both at the surface and at various depths, and soil heat flux. The data from the Campbell tract site was used to derive the relationship parameters, which were then tested against the data taken at the Owens Lake bed.

Results and Discussion. It was found that simple regressions analysis gave unsatisfactory results. This was due to the influence of outlying measurements of low atmospheric heat flux and evaporation whose value was known with little certainty. Error estimates for the calculated z_{0h} and z_{0+} were derived from knowledge of the uncertainties of the measured quantities. With these error estimates, a properly-weighted regressions were performed on two different model relationships between scalar roughness length and z_{0+}. Using the regressed parameters, heat flux from the soil to the atmosphere was predicted and compared to the measured values (Figure 1). It was found that by taking measurement uncertainty into account, the predicted H matched measured H better than theoretical or laboratory based models. It was felt that the difference may arise from the uncertainty about the intermittent nature of surface fluxes and the heterogeneity of the field as compared to lab measurements.

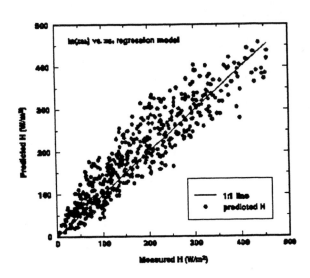

Figure 1. Predicted H vs. measured H for weighted regression model

Literature Cited.

(1) Brutsaert, W. 1975. A theory for local evaporation (or heat transfer) from rough and smooth surfaces at ground level, Water Resour. Res. 11, 543-550.

(2) Brutsaert, W. 1982. Evaporation into the atmosphere}, Kluwer Academic Publishers, Boston.

(3) Albertson, J.D., M.B. Parlange, G.G. Katul, C.R. Chu, H. Stricker and S. Tyler. 1995. Sensible heat flux from arid regions: a simple flux-variance method, Water Resour. Res., 31, 969-974.

(4) Kustas, W.P., B.J. Choudhury, M.S. Moran, R.J. Reginato, R.D. Jackson, I.W. Gay and H.L. Weaver. 1989. Determination of sensible heat flux over sparse canopy using thermal infrared data, Agric. For. Meteorol., 44, 197-216.

Markov Approach to Simulating Geologic Heterogeneity in Alluvial Sediments

Steven F. Carle[1] and Graham E. Fogg[1,2]. [1]*Hydrologic Science, University of California, Davis, CA 95616.* [2]*Dept. of Geology, University of California, Davis, CA 95616.*

Introduction. Gaussian random field assumptions are employed by most stochastic models of heterogeneity in alluvial sediments. Experience shows, however, a number of problems with this approach: subsurface properties are not always normally or log-normally distributed [1]; autocovariance models can seldom be measured or inferred directly from field data, let alone adequately describe geologic patterns; and little or no geologic basis exists for the manner in which the Gaussian model parameters are chosen [2]. Although indicator methods provide the flexibility of a non-parametric approach, the established spatial continuity model development procedures in geostatistics are largely empirical, requiring either abundant data or "training" images [3]. Furthermore, the established indicator assumption of insignificant spatial cross-relationships contradicts geologic observations and, theoretically, causes the order relations problems in the sequential indicator simulation (SIS) algorithm [4] .To address the problem of generating geologically plausible stochastic simulations despite insufficient data coverage, we propose the use of continuous Markov chain models. Such models permit quantification of the geologic concepts that provide a scientific basis for establishing field- to regional-scale heterogeneity patterns in alluvial sediments.

Method. The continuous Markov chain is a mathematically simple, well-understood, yet theoretically powerful probabilistic model. Markov chain models have shown widespread application to one-dimensional (1-D) problems in hydrology and geology, e.g., daily weather occurrences and vertical stratigraphic sequences [5]. We extend Markov chain models to 2- and 3-D by employing Switzer's Theorem, that an n-D Markov chain model exists if 1-D Markov chain models exist in any direction. A 1-D continuous Markov chain model assumes that the probability $p_j(x+h)$ of the occurrence of an event j at a location, say $x+h$, depends entirely on the probability outcomes $p_i(x)$ of the events of the closest location, say x , by

$$p_j(x+h) = p_i(x) \exp[R_{ij}^{(x)} h]$$

where $R_{ij}^{(x)}(h)$ constitutes a transition rate matrix for the x direction. The diagonal rate constants $R_{jj}^{(x)}$ relate to average length $\bar{l}_j^{(x)}$ of an embedded occurrence of j in the x direction by $R_{jj}^{(x)} = -1/\bar{l}_j^{(x)}$.

The off-diagonal transition rates can be interpreted relative to randomness [6]. Such conceptual understanding enables the building of 1-D Markov chain models from stratigraphic interpretations of depositional models. To develop 2- or 3-D continuous Markov chain models, we first develop 1-D models along the major directions of vertical, dip, and strike. We then interpolate the vertical, dip, and strike transition rates to obtain a transition rate matrix for any direction. The conditional simulations are generated in a two-step process. First, the 2- or 3-D Markov chain models are used as basis functions to estimate local probabilities of occurrence in the SIS algorithm, replacing "indicator kriging" estimates. We recognize that the SIS algorithm will not, in theory, produce simulations consistent with two-point spatial continuity models. Thus, a second step of iterative improvement or "simulated quenching" is implemented using the squared deviation from the Markov chain model as the objective function. The resulting simulations honor the spatial structure prescribed by the Markov chain model as well as conditioning data, such as borehole descriptions of sediment types.

Results and Discussion. We have applied this Markov approach to conditionally simulate the spatial distribution of four sediment types, *clay/silt, silty sand, sand/gravel,* and poorly sorted *clay/silt/sand/gravel,* as observed in vertical boreholes drilled into the Arroyo Seco alluvial fan, Livermore, California. We found that a Markov chain model closely fits abundant vertical transition probability measurements obtained from these data. Although the boreholes are relatively closely spaced and numerous, these data remained insufficient to directly model transition probabilities in non-vertical directions. Based on geologic interpretation, the above sediment types can be attributed to *floodplain, overbank, channel* , and *debris flow* deposits, respectively. Geologically plausible average lengths, juxtapositioning patterns, and, consequently, transition rates for dip and strike directions, can largely be inferred from geologic conceptual models. The resulting 3-D Markov chain model was used to generate 3-D conditional simulations. Figure 1 shows a strike section from one such simulation.

Conclusions. Stochastic models of vadose zone properties at the field to regional scale can benefit from integration of geologic concepts to obtain realistic models of heterogeneity. Continuous Markov chains provide a quantitative yet interpretable spatial continuity model that can be developed from data and/or basic concepts of proportion and average length integrated with more subjective geologic concepts of juxtapositioning including cyclicity, asymmetry, and randomness. Once a 2- or 3-D Markov chain model has been developed, it is possible to generate stochastic simulations that honor conditioning data as well as the prescribed spatial continuity model. Such stochastic simulations are useful for evaluating the effects of field-to regional-scale heterogeneity patterns on flow and transport in alluvial sediments.

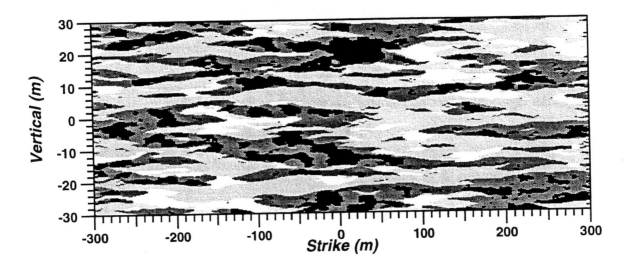

Figure 1. Strike section view of a conditional simulation of sediment architecture in the Arroyo Seco alluvial fan, Livermore, California, with black=*sand/gravel*, dark gray=*silty sand*, light gray=*clay/silt*, and white=*clay/silt/sand/gravel* deposits.

Literature Cited.
(1) Unlu, K., Kavvas, M. L., and Nielsen, D. R., 1989, Stochastic Analysis of Field Measured Unsaturated Hydraulic Conductivity. Water Resources Res., v 25., n. 12, p. 2511-2519.
(2) Neton, M. J., Dorsch, J., Olson, C. D., and Young, S. C., 1994, Architecture and Directional Scales of Heterogeneity in Alluvial Fan Aquifers. Journal of Sedimentary Research, v. 64, n. 2, p. 245-257.
(3) Deutsch, C. V, and Journel, A. J., 1992. Geostatistical Software Library. Oxford Univ. Press., 340 p.
(4) Carle, S. F., and Fogg, G. E., 1995, Transition Probability-Based Indicator Geostatistics. Mathematical Geology, *in press*.
(5) Krumbein, W. C., and Dacey, M. F., 1969, Markov Chains and Embedded Markov Chains In Geology. Mathematical Geology, v. 1, n. 1, p. 79-96.
(6) Miall, A., 1973, Markov Chain Analysis Applied to an Ancient Alluvial Plain Succession. Sedimentology, v. 20, p. 347-364.

Inverse Parameter Estimation for Two-Phase Retention and Conductivity Functions

J. Chen, Y. Liu, J.W. Hopmans, and M.E. Grismer. *Hydrological Science, LAWR, University of California, Davis, CA 95616, USA.*

Introduction. The soil retention and conductivity functions are crucial in quantitative description of subsurface flow and transport problems. Direct measurements of these functions are generally difficult and time consuming. Previous research has demonstrated the feasibility of using the inverse modeling technique to estimate these functions, which involves numerical simulation and parameter optimization based on experiments. Parker et al.[1] was among the first to apply the inverse modeling technique to the one-step outflow method. Eching and Hopmans[2] extended the inverse modeling technique to multi-step outflow experiments of an air-water system. From one-step to multi-step outflow experiments, the computer optimization of the retention curve by inverse modeling was greatly improved. The research described here is the application of the multi-step outflow method to two-phase fluid systems which are air-water, oil (immiscible with water)-water, and air-oil systems. These systems are typical in immiscible organic contaminant research and important in fully understanding subsurface flow transports. The method consists of two parts: transient outflow experiments and the inverse modeling and parameter optimization.

Transient Outflow Experiments. Multi-step outflow experiments were conducted in a laboratory using a modified Tempe cell, which was 7.6 cm high and 6.4 cm in diameter. Colombia sandy loam was air-dried, sieved through a 2-mm screen, and packed uniformly in the cell. Displacement experiments were carried out in three two-phase systems which were (1) air-water, (2) oil (Soltrol 130)-water, and (3) air-oil systems. A soil sample was first saturated with a wetting phase fluid and then drained under a negative suction equivalent to its non-wetting phase entry value. Then multi-steps of positive pressure of the non-wetting phase fluid were introduced to the soil sample, resulting in the wetting phase drainage through a saturated porous plate. Positive pressure steps of non-wetting phase fluids were determined based on the scaled relationships of the interfacial tension between wetting and non-wetting phases. Tensiometers were vertically installed in the middle of the soil samples to monitor pressure changes of wetting phases and the oil non-wetting phase. Transducers, connected to a datalogger, were used for automatic data acquisition of pressures and cumulative outflows during transient flow processes.

Inverse Modeling Approach. The inverse modeling included two parts: the first part was a flow model being the kernel which simulated the transient 1-D two-phase (air-water, air-oil, or oil-water) flows of the experimental systems described above. The second part was an optimization algorithm being the shell which iteratively minimized the difference between the measured and simulated capillary pressures and outflows. The flow model used in this study was a 1-D two-phase finite element model with a modified Picard linearization. The modified Picard linearization was based on the mixed form of the unsaturated flow governing equation and was inherently mass conservative[3]. The optimization was formulated by a Weighted Least Squares (WLS) problem and solved by a Levenberg-Marquardt (LM) method[4]. The LM solution algorithm was commonly used in practices of nonlinear least square problems.

The objective of the inverse parameter estimation was to determine the soil retention and conductivity function parameters, which were, in turn, the constitutive relationships of the inverse modeling. From the available parametric models of these functions, van Genuchten-Mualem model was chosen in this study. Five parameters (θ_r, α, n, K_{sw} - wetting phase K_s, and K_{snw} - non-wetting phase K_s) of both retention and conductivity functions were simultaneously estimated in the optimization. The uniqueness of the proposed parameter estimation method was examined by using different initial parameter guesses, and the parameter sensitivities were evaluated by using different parameter combinations. Scaling relationships in two-phase flow systems were used as a cross validation of the derived results.

Results and Discussion. In the original multi-step outflow experiments[2] of an air-water system, air pressure was assumed to be constant in the system during experiments, because of its low viscosity and density. When a non-wetting phase was oil in an oil-water system, this assumption was no longer valid, since the viscosity and density of the oil phase were in the same order of magnitude of the wetting phase, water. Therefore hydrophobic tensiometers were used to monitor positive pressure changes of the oil phase in the oil-water system. The measurements showed that the positive oil pressure in the cell reached relatively constant values in much less time than that of water, in other words, the pressure gradient of the oil phase was very small during drainage period of the wetting phase fluid. These results were consistent with a numerical simulation of the two phase flow system.

It was observed that the entry value of the non-wetting phase was the smallest for an air-oil system and the largest for an air-water system. This observation could be explained by the interfacial tension between wetting phase and non-wetting phase for each system. The parameters α and n of the van Genuchten retention function were inversely related to the entry value of non-wetting phase and the width of the pore size distribution, respectively. Therefore, different α values and a constant n value were expected. It was found that among optimized parameters, α was the smallest for an air-water system, while n was approximately constant for the three two-phase systems (Table 1).

Based on the outflow experimental data, the inverse parameter optimization proved to be convergent and unique for different initial estimates and had an excellent sensitivity for the retention function parameter estimation. The results also showed a relatively low sensitivity for the non-wetting phase conductivities, which will be shown to be caused by the physics of the experimental system. The presented method is specially useful for determination of two phase retention and conductivity functions of soil.

Table 1. Optimized Parameters of van Genuchten Equation for Columbia Sandy Loam

	Air-Water	Oil-Water	Air-Oil
θ_r (cm^3/cm^3)	0.123	0.098	0.053
K_{sw} (cm/hr)	3.0	8.9	2.1
K_{snw} (cm/hr)	0.2	0.5	0.3
α (1/cm)	0.011	0.021	0.029
n	2.23	2.68	2.54
θ_s^* (cm^3/cm^3)	0.464	0.464	0.464

* fixed value

Literature Cited.
(1)   Parker, J.C., J.B. Kool, and M. Th. van Genuchten. 1985. Determining soil hydraulic properties from one-step outflow experiments by parameter estimation: II. experiment studies. Soil Sci. Soc. Am. J. 49:1354-1359.
(2)   Eching, S.O. and J.W. Hopmans. 1993. Optimization of hydraulic functions from transient outflow and soil water pressure data. Soil Sci. Soc. Am. J. 57:1167-1175.
(3)   Celia, M.A. and E.T. Bouloutas. 1990. A general mass-conservative numerical solution for the unsaturated flow equation Water Resour. Res. 26:1483-1496.
(4)   Kool, J.B. and J.C. Parker. 1988. Analysis of the inverse problem for transient unsaturated flow. Water Resour. Res. 24:817-830.

Noninvasive Observation of Pore-Scale Transport And Plant-Roots With X-Ray Computed Tomography

V. Clausnitzer, D. A. Heeraman, J. W. Hopmans, and J. S. Steude. *Hydrologic Science, Department of Land, Air & Water Resources, University of California, Davis, CA, 95616, U.S.A.*

Introduction. The study of water movement, fate and transport of chemicals, and plant-root activity in soils is of fundamental importance in hydrologic science. Contaminants migrating through the subsurface constitute an ongoing threat to natural environments, groundwater quality, and public health. Root systems and their dynamic responses in growth and uptake to various soil conditions can profoundly affect various vadose-zone processes. Current concepts and models are only partly successful in predicting chemical transport and plant-soil interaction. It becomes increasingly clear that in order to improve our predictive abilities we need to better account for the fundamental physical processes controlling fluid behavior at the pore scale, as well as a better understanding of the microscale interrelationships of soil and root characteristics.

Traditional measurement techniques are invasive and can substantially disturb the very property to be measured in the vicinity of the sampling location. They typically average physical and chemical parameters over relatively large soil volumes and cannot discern inhomogeneities within the sampled volume. The methods for studying root systems are often destructive, tedious, and difficult to interpret (Bohm, 1979; Russell, 1977; Schuurman and Goedewaagen, 1971).

X-ray computed tomography (CT) can noninvasively measure soil properties at a scale much smaller than conventional sampling methods (Hopmans et al., 1994). The technique has been applied in the earth sciences since the mid-1980's. CT output is directly related to density and atomic number and thus can be used for phase identification (including plant roots) and concentration measurements. We report on the potential of x-ray microtomography for studying (1) multiphase and solute transport in porous media at 15 μm resolution and (2) the spatial distribution of plant roots in situ at 200 μm resolution.

Materials & Methods. All scans were performed at Scientific Measurement Systems (SMS), Inc., (Austin, Tx.). A set of computational procedures has been developed to visualize and interpret the volumetric data sets of attenuation values that are produced by the scanning techniques used at SMS. Frequency histograms were obtained for all 3D-data sets to separate attenuation-value ranges for each individual component in the respective sample.

Transport Study: 50-mm glass spheres were used as unconsolidated porous material in a flow cell of 5 mm inner diameter. We first scanned the glass-sphere pack and different fluid-phase combinations (CCl_4, H_2O, Air) within the pore space under static conditions. Recent experiments involved the study of dynamic systems such as finger development during the displacement of water from a porous medium by oil, and the breakthrough of a NaI-solution pulse in an initially water-saturated sample. The flow rate was 100 ml/hr in both cases. Using a 10-μm diameter microfocus x-ray source (cone beam) and a planar detector composed of 20-μm × 20-μm cells resulted in a final image resolution of approximately 15 μm. Sets of 30 horizontal slices covering approximately 0.3 mm vertical range of the flow cell were acquired simultaneously.

Root Study: Bush bean (<u>Phaseolus</u> <u>vulgaris</u> L.) seedlings were grown under optimum conditions for two weeks in individual PVC columns (10 cm long, 5 cm inner diameter) using an Oso-Flaco fine sand soil medium. The stems of the plants were excised at their base and the root systems imaged with a high-energy scanner using a fan-beam source and a linear detector. Each horizontal tomogram represents a 412 × 412 matrix of relative-attenuation values for individual 0.2 mm × 0.2 mm × 0.2 mm sample voxels. 3D images were obtained after combining 40 individual tomograms into a single set representing an 8 mm high section of the scanned sample. For reference, several plant-root systems were destructively sampled and analyzed for total root length, root diameter and dry weight.

Results & Discussion

Transport Study: We first obtained 3D pore-space representations for the glass-sphere packs and were able to identify individual fluid phases (CCl_4, H_2O, Air) within the pore space under static conditions. Fig. 1 shows a single horizontal tomogram for one such case. We determined that the finite acquisition time of 20 min for each set of 30 horizontal tomograms did not cause artifacts in the dynamic cases. We present results for both the static and dynamic cases, including pore-scale breakthrough curves, that illustrate the visualization and quantitative capabilities of CT when combined with suitable data processing.

Root Study: The line graph plots indicate the approximate location of the various column materials in the tomogram and upper and lower limits of relative attenuation classes for the soil and root inside the column. Volume rendering of the data set qualitatively displayed the spatial arrangement of plant roots with diameters of 200 μm and larger in the soil matrix. The combined data set showed a single-peak frequency distribution indicating that substantial noise was introduced to the system due to partial-volume effects involving pore air, pore water, solid particles, and roots. We will supplement qualitative data with quantitative estimates of total root length, root volume and root length per soil surface area by estimating the relative fractions of air, root and matrix within each voxel. Comparisons will be made with rooting parameters obtained from destructive sampling.

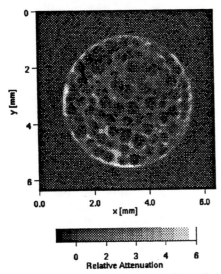

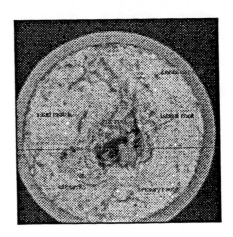

Fig. 1 CCl_4-ganglia after image subtraction "dry" from "wet"

Fig. 2   Horizontal Tomogram showing individual roots (5 cm I.D.)

References

(1) Bohm, W. 1979. Methods of studying root systems. Springer-Verlag, New York, Inc., New York. Hopmans, J.W., M. Cislerova and T. Vogel. 1994. X-ray tomography of soil properties, in: Tomography of soil-water-root processes, Soil Sci. Soc. of America, Special Publ. No. 36, pp. 17-28.

(2) Russell R S 1977 Plant Root Systems: Their Function and Interaction with Soil. McGraw-Hill, London. 298 p.

(3) Schuurman, J.J., and Goedewaagen, M.A.J. 1971. Methods for the examination of root systems and roots. Centre for Agricultural Publications and Documentation; Wageninigen, The Netherlands.

In Situ Measurement of the Linear Shrinkage Curve of Weakly Swelling Soils

Y. Coquet[1], P. Boivin[1], E. Braudeau[1], and J. Touma[2]. [1]Institut Français de Recherche Scientifique pour le Développement en Coopération (ORSTOM), B.P. 1386, Dakar, Senegal. [2]ORSTOM, B.P. 5045, 34032 Montpellier Cedex 1, France.

Introduction. The linear shrinkage curve (LSC) of a soil expresses its relative linear deformation ($\Delta L/L_0$) as a function of its gravimetric water content, and most generally, is determined in the laboratory on undisturbed samples : either on clods by volumetric measurements and assuming isotropic deformation (1), or on cores by assessing directly the linear shrinkage with Vernier calipers (2) or some other measuring device such as displacement transducers (3). These last permit very precise measurements ($\pm 1\mu m$) under isothermal conditions, and a quasi-continuous LSC is obtained. The general use of this procedure showed that the sigmoidal shape of the shrinkage curve could be regarded as a very general property of soils, including very weakly swelling soils. In situ determination of the LSC has been seldom attempted, and mostly on highly swelling soils (4,5). Some authors (6) showed that the in situ LSC of soil layers could differ from the laboratory LSC of samples taken from those layers. In this paper we present an attempt of in situ measurement of the LSC of a weakly swelling soil and compare the results to soil core LSCs.

Materials and Methods. The in situ LSC of a Loamy Aquic Torrifluvent (Eutric Fluvisol) was obtained after wetting by a water pond. We used a set of shrink-swell gauges to measure the thickness variations of soil layers. Each gauge is composed of a rod anchored into the soil by its threaded bottom end. The gauges were placed 10 cm beside each other, at depths multiple of 20 cm. The displacement of the top of the rods was measured by electronic displacement transducers relative to a single rigid plate laying on the topsoil. The gravimetric water content of the soil layers being impossible to assess on the same verticals as the thickness measurements, was inferred from samples taken around the shrink-swell gauges by block kriging interpolation. At the end of the experiment, 5 cm-diameter/height undisturbed cores were extracted from the soil layers and their LSCs measured after rewetting in the laboratory.

Results and Discussion. Provided a correction for temperature effects on the displacement transducers, the thickness measurement error was at the maximum $\pm$ 10 μm. The water content precision (95 % confidence interval) was in average $\pm$ 0.015 g/g (ranging from 0.010 to 0.019 g/g). The shape of the in situ LSCs does not depart from the general description of the LSCs (7). Starting from the highest water content, one can distinguish a structural shrinkage stage corresponding to a slight deformation for a large variation of water content (Fig. 1). The slope of the LSC inceases progressively until reaching the start of a basic shrinkage stage (7). Because of the limited water content range encountered in situ, especially for the deepest layers, one can only estimate a minor of the slope of the basic shrinkage stage.

The in situ LSCs were compared to the laboratory LSCs measured on cores (Fig.1). The initial dimension L_0 of the soil layers is that just after the installation of the shrink-swell gauges, before wetting the soil,while the reference dimension L_0 of the soil cores is that corresponding to the same initial water content W_0 measured in situ. The agreement between in situ and laboratory LSCs is good, compared to the differences observed between the LSCs of the different soil layers. Although, two systematic differences between in situ and laboratory LSCs appear. The first is that the latter overestimates the slope of the in situ structural shrinkage stage (table 1), particularly for the top layer. The second divergence is that the basic shrinkage stage of the in situ LSCs is attained at a water content increasing with depth. This results in an increased shift of the in situ LSC towards high water content, when compared to the laboratory LSC (Table 1). These effects may be imputable to the differences in overburden and/or confining pressure conditions between in situ and laboratory determinations. The steep slopes observed at higher water contents in laboratory LSCs, especially for the 0-20 and 20-40cm layers, are likely due to a collapse of the cores at the beginning of shrinkage.

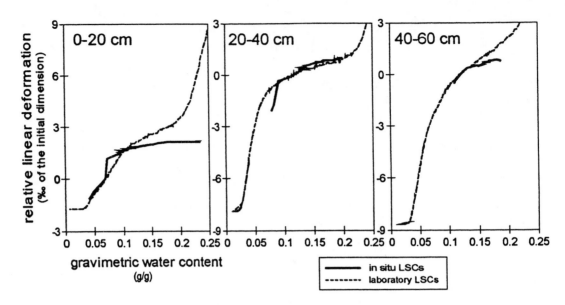

Fig. 1. Linear shrinkage curves (LSCs) of three soil layers measured in situ (continous lines) or in the laboratory on 100 cm³ cores extracted from those layers (dashed lines).

depth (cm)	type of measurement	slope of structural shrinkage stage (mm/mm.(g/g)⁻¹)	slope of basic shrinkage stage (mm/mm.(g/g)⁻¹)	water content at structural/basic transition(g/g)
0 - 20	in situ	.0027	.07	.080
	laboratory	.0150	.05	.105
20 - 40	in situ	.0073	.13	.096
	laboratory	.0100	.21	.063
40-60	in situ	.0072	.10	.118
	laboratory	.0250	.18	.075

Table 1. LSC parameters estimated from in situ or laboratory measurements for the three layers.

Literature Cited.
(1)   Grossman R.B., Brasher B.R., Franzmeier D.P., and Walker J.L. 1968. Linear extensibility as calculated from natural-clod bulk density measurements. Soil Sci. Soc. Am. Proc., 32 : 570-573.

(2)   Towner G.D. 1986. Anisotropic shrinkage of clay cores, and the interpretation of field observations of vertical soil movement. J. Soil Sci. 37 : 363-371.

(3)   Braudeau E. 1987. Mesure automatique de la rtraction d'chantillons de sol non remanéis. Science du Sol. 25 : 85-93.

(4)   Bozozuk M., and Burn K.N. 1960. Vertical ground movements near elm trees. Geotechnique. 10 : 19-32.

(5)   Mitchell A.R., and Van Genuchten M.T. 1992. Shrinkage of bare and cultivated soils. Soil Sci. Soc. Am. J. 56 : 1036-1042.

(6)   Hallaire V. 1987. Retrait vertical d'un sol argileux au cours du desschement. Mesures de l'affaissement et conséquences structurales . Agronomie. 7 : 631-637.

(7)   Mitchell A.R. 1992. Shrinkage terminology : escape from "normalcy". Soil Sci. Soc. Am. J., 56 :993-994.

2-D and 3-D Wetting Front Instabilities in Layered Soils Investigated Through Image Techniques and Invasion Percolation Model

S. Crestana and A. N. D. Posadas. *Centro Nacional de Pesquisa e Desenvolvimento de Instrumentação Agropecuária, CNPDIA - EMBRAPA, Cx.P. 741, CEP 13560-970, São Carlos, SP., Brazil (work supported by FAPESP, Proj. nº 90/3773-7).*

Introduction. In recent years, great attention has been given to the study of fluid transport in porous media. Its main reasons are the economic interest in improving the secondary and tertiary recovery of petroleum and the possible water and soil contamination due to intensive use of agrochemicals. In order to understand this extremely complex problem, some researchers have devoted themselves to the study of fluid transport in connection to the geometry of porous media (1). Many experiments done in this way have shown that the fluid transport-porous medium coupling has auto-similarity or fractal characteristics in a range of defined scales (2). Those characteristics have led to the creation of some simulation models such as the Diffusion Limited Aggregation-DLA which simulates the viscous fingering phenomenon (3) and the Percolation through Invasion (4), which simulates the capillary fingering phenomenon. The phenomenon of fingering in soils, which is basically of capillary character, also presents fractal characteristics (5,6), making it clear that the use of the modified theory of percolation through invasion is an appropriate model of simulation (7). Thus, the purpose of this paper is to introduce the image processing, X-ray and nuclear magnetic resonance tomography for the characterization of the fingering phenomena in soils, as well as the aplication of the fractal theory together with a modified invasion percolation model to describe the fingers (7).

Materials and Methods. The physical process concerning to the fingering phenomena in two and three dimensions experiments of water infiltration through layered soil columns were carried out in the laboratory. In two dimensions the fingering dynamics through a planar column of two-layer soil was followed by using imaging techniques. The fractal theory together with a modified invasion percolation model have been applied to describe the fingering phenomena in soils (7). In three dimensions, the infiltration of water trhough a cubic column of two-layer soil was studied under hydrodynamic steady state conditions using images obtained by magnetic resonance (MR). The fingering dynamics into a vertical slice located in the center of the cubic column was studied taking into account images obtained by medical X-ray tomography. Also, it was possible to follow the spatial dynamics of the wetting front by means of a MR system measuring the spin echo signal (8).

Results and Discussion. The results showed that the perimeter of the fingers displacement front presents a behavior of self-affine fractal, being its fractal dimension D_p dependent on the time and the position along the profile of the soil column. On the other hand, it was verified, by means of the several experiments carried out, that the fractal characteristics of the fingers, under the same initial conditions, are related to the soil granulometry or the pore profile distribution. For fine textured soils the structure of the fingers is non-fractal, whereas for coarse textured soils the structure of the fingers present statistically self-affine or fractal characteristic. In addition to these results, it was also possible to introduce the modified theory of percolation by invasion, so as to morphologically simulate the fingers structure (Fig.1) and the pore profile distribution (7). The employed tomography and imaging techniques already available, through nuclear magnetic resonance as well as through X-ray (Fig.2), presented very innovating and encouraging results for the study of the fingering phenomenon in soils and porous media in general, allowing us, among other things, to witness, in a non-disturbing way, the three-dimensional and aleatory character of the phenomenon. Nevertheless, great research challenges remain. It is needed, for instance, to better relate the "effective" surface tension to the fractality of the fingers (5), the passage from non-fractality to the fractal behavior of the finger structure in relation to the soil granulometry, and the pore profile distribution to the phenomenon occurrence (7). Concomitantly, the exploration and the development of more appropriate methods to measure their morphological and physical characteristics, in 2 and 3 dimensions, establish an important challenge to be faced in the near future.

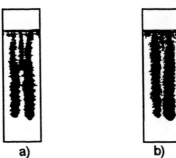

a)   b)

Fig. 1. a) Processed image for the experimental result of the fingering at the end of the experiment. b) Typical cluster obtained by simulation through the program of modified invasion percolation.

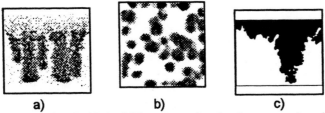

a)   b)   c)

Fig. 2. a) and b) Images obtained with the MR system, showing transversal and coronal sections of the fingering phenomenon in static conditions. c) Processed image obtained through a medical X-ray tomograph, representing the growth of fingers through the central transversal section of a cubic column of soil.

Literature Cited

(1)   Lu, T.X., Biggar, J.W., and Nielsen, D.R. 1994. Water movement in glass bead porous media. 2. Experiments of infiltration and finger flow. Water Res. Res. 30(12): 3283-3290.

(2)   Katz, A.J. and Thompson, A.H. 1985. Fractal sandstone pores: implications for conductivity and pore formation. Phys. Rev. Lett. 54(12):1325-1328.

(3)   Chen, J.D. and Wilkinson, D. 1985. Pore-scale viscous fingering in porous media. Phys. Rev. Lett. 58(18): 1892-1895.

(4)   Wilkinson, D., and Willemsen, J.F. 1983. Invasion Percolation: a new form for percolation theory. J. Phys. A: Math. Gen. 16: 3365-3376.

(5)   Chang, W.-L., Biggar, J.W., and Nielsen, D.R. 1994. Fractal description of wetting front instability in layered soils. Water Res. Res. 30(1): 125-132.

(6)   Posadas, D.A.N., and Crestana, S. 1993. Aplicações da teoria fractal na caracterização do fenômeno "fingering" em solos não saturados. Rev. Bras. Ci. Solo. 17(1): 1-8.

(7)   Onody, R.N., Posadas, D.A.N., and Crestana, S. 1995. Experimental studies of fingering phenomena in two dimensions and simulation using a modified invasion percolation. J. of Appl. Phys. In press.

(8)   Posadas, D.A.N., Tannús, A. Panepucci, C.H., and Crestana, S. 1995. Magnetic resonance imaging as a non-invasive technique for investigating 3-D preferential flow occurring within stratified soil samples. Computers and Electronocs in Agriculture (submitted to).

In Situ Determination of Soil Hydraulic Properties In a Large-Scale Field Experiment

R. H. Cuenca, D. E. Stangel and S. F. Kelly. *Department of Bioresource Engineering, Gilmore Hall, Oregon State University, Corvallis, OR 97331-3906. e-mail: cuencarh@pandora.bre.orst.edu*

Introduction. The boreal ecosystem-atmospheric study (BOREAS) is a large-scale atmospheric-hydrologic-ecosystem-remote sensing experiment conducted in the boreal forest of Canada. Major field campaigns were conducted in 1994. Measurements of soil status and soil physical properties at each flux tower site included monitoring profiles of soil water content and soil water potential as well as tension infiltrometer tests. One of the objectives for our team was to determine *in situ* soil hydraulic properties such as soil water retention and unsaturated hydraulic conductivity functions which could be applied in simulation models for various components of the boreal forest ecosystem. The resulting functions will be submitted to the BOREAS Information System (BORIS) for use by modelers and other research scientists.

$$h(\theta_v) = \frac{1}{\alpha}[(S_e)^{-1/m} - 1]^{1/n} \qquad [1]$$

Materials and Methods. Instrumentation applied for soil profile measurements included neutron probe, time domain reflectometry (TDR) and tension blocks. The tension blocks were installed in a course sand soil and the readings were not felt to be reliable, probably due to limited contact area. Data analysis for unsaturated hydraulic conductivity reported in this abstract were taken from a sandy soil site in which soil water content was measured by neutron probe. The frequency of measurement was every other day and readings were taken at 10 cm intervals starting at a depth of 5 cm. Readings taken at depths less than 20 cm were adjusted for neutron escape using the method of Parkes and Siam (1979). Data management, computation and graphic display of soil water content and tension profiles were accomplished using the HYDROSOL computer program. Tension infiltrometers of two radii operated at three tensions were used to evaluate the hydraulic conductivity function. The basic equations fit to the data were the van Genuchten soil water retention function (Eq. 1) and hydraulic conductivity function using the constraint of Mualem (Eq. 2) given as:

$$K(S_e) = K_s S_e^{l}[1 - (1 - s_e^{1/m})^m]^2 \qquad [2]$$

where $h()$ = soil water tension as function of soil water content (cm), = 1/air entry pressure (1/cm), m, n = fitting parameters, $K(S_e)$ = hydraulic conductivity as function of effective saturation (cm/d), K_s = saturated hydraulic conductivity (cm/d), l = tortuosity (fitting parameter), $m = 1 - 1/n$, and S_e = effective saturation (fraction) given as:

$$S_e = \frac{\theta_v - \theta_r}{\theta_s - \theta_r} \qquad [3]$$

where $_v$ = volumetric soil water content (cm^3/cm^3), $_r$ = residual volumetric water content (cm^3/cm^3), $_s$ = saturated volumetric water content (cm^3/cm^3) (van Genuchten, 1980).

The flux-gradient relationship in a Darcian flow field applies the hydraulic conductivity function to quantify the relationship between the mass flux and total head gradient. Soil water profiles measured in BOREAS were analyzed with an appropriate soil water retention function to plot the total head profile and locate the zero-flux plane separating evapotranspiration moving in response to an upward gradient from drainage moving down through the soil profile. The time series of soil water profiles was applied to compute the mass flux between any two time intervals for all depths below the zero-flux plane in the profile. Application of a soil water retention function allowed for determination of the total head gradient and solution of the *in situ* unsaturated hydraulic soil water retention function allowed for determination of the total head gradient and solution of the *in situ* unsaturated hydraulic conductivity. Data had to be screened to minimize the effects of instrument error from the neutron probe. Only time intervals and depth layers in which significant fluxes were observed were retained in the solution set for hydraulic conductivity. The hydraulic conductivity computed using the flux-gradient data at depth was combined with that determined by the tension infiltrometer on the soil surface and the fitting parameters of the soil water retention function were optimized to minimize the estimation error.

Results and Discussion. Results from tension infiltrometer tests and flux gradient analysis conducted for the same site were used to derive the unsaturated hydraulic conductivity function shown in Fig. 1. As shown in the figure, the tension infiltrometer data were used to define the moist part of the function and specifically the value of saturated hydraulic conductivity. All of the screened data were used to produce the data shown in the portion of the figure labeled Flux-Gradient Data. Curve fitting techniques applied separately to the tension infiltrometer data or flux-gradient data proved inadequate for defining the unsaturated hydraulic conductivity function over the range of tension shown in the figure. Only by combining the two data sets could the fitting parameters of the van Genuchten function be determined over the range of tension which would be useful for defining the soil hydraulic properties for this site and reproducing by numerical simulation model the measured drainage profiles. The parameters fit to *in situ* data were significantly different from those reported in the literature by Carsel and Parrish (1988) and Rawls et al. (1982) and felt to be more representative of the field conditions.

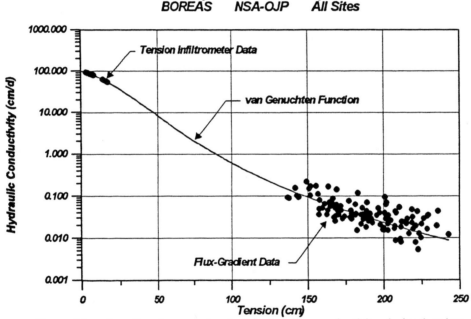

Fig. 1 van Genuchten function for unsaturated hydraulic conductivity derived using combined tension infiltrometer and flux-gradient data, site NSA-OJP, BOREAS experiment.

Literature Cited.

(1) van Genuchten, M. Th. 1980. A closed-form equation for predicting the hydraulic conductivity of unsaturated soils. *Soil Science Society of American Journal*, 44, 892-898.

(2) Parkes, M. E. and Siam, N. 1979. Error associated with measurement of soil moisture change by neutron probe. *Journal of Agricultural Engineering Research*, vol 24, pp. 87-93.

(3) Carsel, R. F. and R. S. Parrish. 1988. Developing joint probability distributions of soil water retention characteristics. *Water Resources Research*, vol. 24, pp. 755-769.

(4) Rawls, W. J., D. L. Brakensiek and K. E. Saxton. 1982. Estimating soil water properties, *Transactions, American Society of Agricultural Engineers*, vol. 25, No. 5, pp. 1316-1320, 1328.

The Potential for Modelling Wetting Front Instability Using a Geotechnical Centrifuge

P.J. Culligan-Hensley[1], D.A. Barry[2], and J.-Y. Parlange[3]. [1]*Massachusetts Institute of Technology, Cambridge, MA 02139, USA.* [2]*University of Western Australia, Nedlands, WA 6907, Australia. Cornell University, Ithaca, NY 14853, USA.*

Introduction. Geotechnical centrifuge modelling is a technique that has proved useful in the study of one-dimensional stable, wetting displacements in unsaturated media (1). Under certain conditions, however, wetting fronts infiltrating unsaturated media can become unstable and break into fingers, which move vertically to the water table, bypassing a large proportion of the vadose zone. Immiscible transport under these circumstances is extremely non-uniform. As a result, fluid loading at the water table will be quite different than if transport is assumed to be one-dimensional. For this reason, contaminant hydrologists are paying increasing attention to the subject of wetting front instability. This abstract addresses the potential for centrifuge modelling of unstable flow in porous media, with particular reference to the phenomenon of gravity-driven wetting front instability.

Principle of Centrifuge Modelling. Geotechnical centrifuge modelling is an experimental method traditionally used by geotechnical engineers to obtain soil conditions that are homologous in model and prototype (2). This is achieved by subjecting a scale soil model, where all linear dimensions are reduced by a factor n, to a centrifugal acceleration of n gravities (ng) (Fig. 1). The application of centrifugation in soil science is also well established: over 20 years ago Alemi, Nielsen and Biggar demonstrated the experimental determination of transport parameters in soil cores using centrifugal techniques (3). More recently, research work in the field of environmental engineering has made use of the method to investigate the behaviour of gravity-driven flow phenomena under realistic, but well-controlled, field boundary conditions (4).

General scaling relationships for flow phenomena in the centrifuge have been developed by numerous authors (5). For a reduced-scale centrifuge model test conducted using prototype soil and fluids, the relationships given in Table 1 are either evident or well established. In what follows, it will be assumed that these relationships are also applicable to the centrifuge modelling of wetting front instability.

Table 1: Centrifuge Scaling Relationships

Parameter	Prototype/model ratio
Gravity, g	1/n
Macroscopic length, L	n
Microscopic length, d	1
Pore fluid velocity, u	1/n
Seepage flux, q	1/n
Fluid pressure, p	1
Fluid pressure head, ψ	n
Time, t	n^2
Hydraulic conductivity, K	1/n
Soil permeability, k	1
Saturated fluid content, θ	1
Diffusivity, D	1
Fluid density, ρ	1
Fluid viscosity, μ	1

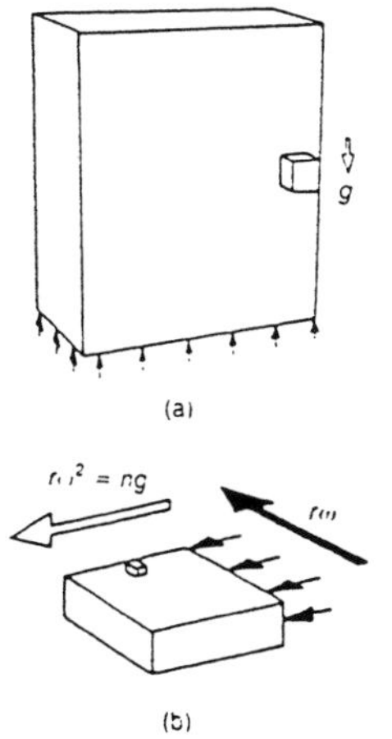

Fig 1. Gravity effects in prototype and model (2)
(a) prototype; (b) model

Wetting Front Instability. The important factors used to characterise unstable wetting displacements are (i) the conditions for the onset of fingering, (ii) the finger width and (iii) the finger velocity.

Conditions for the onset of wetting front instability are generally derived from linear analyses. These consider the stability of a small perturbation at the macroscopic interface separating the two immiscible fluids. For two-dimensional disturbances, the onset of fingering is associated with a perturbation containing wavelengths (tip-to-tip separations) greater than a critical wavelength λ_c. The average finger width, w, is usually obtained by identifying the most rapidly growing wavelength. By performing dimensional analysis at the finger scale, the finger velocity, u_f, can also be obtained (7).

In what follows, we shall adopt the following expressions for λ_c, w and u_f (6, 7, 8):

$$\lambda_c = \frac{2\pi\Gamma(\mu_1+\mu_2)\theta_F}{u\theta_F(\mu_2 - \mu_1)+k_F g(\rho_1-\rho_2)} \tag{1}$$

where u is the velocity of an unperturbed flat front, and where Γ is described by

$$\Gamma = \int_{\psi_0}^{\psi_F} \frac{K d\psi}{\theta_F-\theta_0} \tag{2}$$

where K is the hydraulic conductivity, ψ is the fluid pressure head, and the subscripts 0 and F denote, respectively, values at the front and in back of a capillary-induced diffusion zone at the infiltrating front;

$$w = \frac{S_F^2}{K_F(\theta_F - \theta_0)} f_{dF}(R_F) \tag{3}$$

where f_{dF} is a function of R_F, the dimensionless flux-conductivity ratio $(R_F = q/K)$ and where S is the sorptivity in back of the diffuse zone, which can be approximated by (9)

$$S_F^2 = \int_{\theta_0}^{\theta_F} (\theta + \theta_F - 2\theta_0)D\, d\theta \tag{4}$$

$$u_f = \frac{K_F}{\theta_F - \theta_0} f_{vF}(R_F) \tag{5}$$

where f_{vF} is also a function of R_F.

Centrifuge Scaling Ratios. Equations 1 - 5 can be used to derive scaling ratios relevant to the centrifuge modelling of unstable wetting displacements.

For similitude between model and prototype with respect to the conditions for the onset of fingering, the critical wavelength, λ_c, which is a macroscopic length, should be n times higher in the prototype than the centrifuge model (refer to Table 1); in other words, $\lambda_{c(r)} \equiv n$, where r denotes the ratio between a value in the prototype to that in the model (i.e. $\lambda_{c(r)} \equiv \lambda_{c(p)}/\lambda_{c(m)}$). This criterion ensures that the initial wavelength (tip-to-tip separation) of disturbances observed during the onset of fingering is n times smaller in a centrifuge model than in the corresponding prototype.

28

Through inspectoral analysis, we can use the definition of λ_c given by Equation 1 to assert that

$$\lambda_{c(r)}\lambda_{c(m)} \equiv \frac{2\pi\Gamma_r\Gamma_m(\mu_{1(r)}\mu_{1(m)}+\mu_{2(r)}\mu_{2(m)})\theta_{F(r)}\theta_{F(m)}}{u_r\mu_m\theta_{F(r)}\theta_{F(m)}(\mu_{2(r)}\mu_{2(m)}-\mu_{1(r)}\mu_{1(m)})+k_{F(r)}k_{F(m)}g_rg_m(\rho_{1(r)}\rho_{1(m)}-\rho_{2(r)}}$$ [6]

From Table 1, $g_r = \psi = 1/n$ and $\mu = \rho = 1$. In addition, experimental evidence indicates that θ is approximately equal to θ_s, the saturated value of the infiltrating fluid content (7). Hence, it is reasonable to assume that $\theta_{F(r)} = 1$, which, in turn, suggests that $k_{F(r)} = 1$. Provided that the model-preparation procedure fully accounts for the wetting history of the prototype under investigation, $\theta_{0(r)} = 1$. Thus, from Equation 2, $\Gamma_r = 1$.

Substitution of all relevant ratios into Equation 6 leads to the conclusion that

$$\lambda_{c(p)} \equiv \frac{1}{1/n}\lambda_{c(m)} = n\,\lambda_{c(m)}$$ [7]

which satisfies the criteria for similitude between model and prototype with respect to the conditions for the onset of fingering.

For the correct scaling of average finger width between model and prototype, it is also necessary that $w_r = n$. In other words, it is necessary that the width of fingers formed in a centrifuge model is n times smaller than those formed in the prototype. Again, this criterion arises because finger width is a macroscopic length.

Using the approximation for finger width given by Equation 3, we may ascertain that

$$w_p \equiv w_r w_m \equiv \frac{S_{F(r)}^2 S_{F(m)}^2}{K_{F(r)}K_{F(m)}(\theta_{F(r)}\theta_{F(m)} - \theta_{0(r)}\theta_{0(m)})}[f_{dF}(R_F)]_r\,[f_{dF}(R_F)]_m$$ [8]

where, from Equation 4,

$$S_{F(r)}^2 \equiv \theta_r D_r = 1$$ [9]

and where $[f_{dF}(R_F)]_r = 1$, because R_F, the argument of the function, is the same in model and prototype ($R_{F(r)} \equiv q_{F(r)}/K_{F(r)} = 1$).

By substituting all applicable ratios into Equation 8, we find that

$$w_p \equiv \frac{1}{1/n}w_m = n\,w_m$$ [10]

Thus, the correct scaling of finger width may be achieved during reduced-scale centrifuge modelling.

Finally, for similitude, the scaling laws for flow phenomena require that flow velocities in a centrifuge model be n times higher than those in the corresponding prototype (refer to Table 1). Thus, it is necessary that the finger velocity u_f scales as $u_{f(r)} = 1/n$.

The finger-propagation velocity was described by Equation 5, which can be used to show that

$$u_{f(p)} \equiv u_{f(r)}u_{f(m)} \equiv \frac{K_{F(r)}K_{F(m)}}{\theta_{F(r)}\theta_{F(m)} - \theta_{0(r)}\theta_{0(m)}}[f_{vF}(R_F)]_r\,[f_{vF}(R_F)]_m$$ [11]

Substitution of the appropriate ratios in Equation 11 leads us to conclude that

$$u_{f(p)} = \frac{1}{n} u_{f(m)}$$

[12]

Hence, the finger propagation velocity also scales in accordance with the general requirement for the centrifuge scaling of fluid velocity.

Discussion.   The scaling ratios presented above theoretically demonstrate the feasibility of modelling unstable wetting displacements using a geotechnical centrifuge. However, as in all cases, the applicability of these ratios will be subject to certain conditions. Experimental studies are currently underway at Massachusetts Institute of Technology and the University of Western Australia to confirm the feasibility of modelling wetting front instability using a geotechnical centrifuge. To date, the data from these studies are promising; they suggest that a geotechnical centrifuge offers a useful tool for studying unstable finger flow under realistic field-boundary conditions.

Literature Cited.
(1)   Illangasekare, T.H., Znidarčić, Al-Sheridda, M. and Reible, D.D. 1991. Multiphase flow in porous media, Centrifuge 91, H.Y. Ko and F.G. McLean (eds.), 517-523.
(2)   Schofield, A.N. 1980. Cambridge geotechnical centrifuge operations, Twentieth Rankine Lecture. Geotechnique, 30(3), 227-268.
(3)   Alemi, M.H., Nielsen, D.R., and Biggar, J.W. 1976. Determining the hydraulic conductivity of soil cores by centrifugation, Soil Sci. Soc. Am. J., Vol. 40, 212-218.
(4)   Hensley, P.J. and Savvidou, C.S. 1993. Modelling coupled heat and contaminant transport in groundwater, International Journal for Numerical and Analytical Methods in Geomechanics, 17, 493-527.
(5)   Culligan-Hensley, P.J. and Savvidou, C. 1994. Environmental geomechanics and transport processes, in Geotechnical Centrifuge Testing, Ed. R.N. Taylor, Blackie Academic.
(6)   Parlange, J.-Y. and Hill, D.E. 1976. Theoretical analysis of wetting front instability in soils, Soil Sci., 122, 236-239.
(7)   Glass, R.J., Parlange, J.-Y. and Steenhuis, T.S. 1989. Wetting front instability, 1, Theoretical discussion and dimensional analysis, Water Resour. Res., 25, 1187-1194.
(8)   Glass, R.J., Parlange, J.-Y. and Steenhuis, T.S. 1991. Immiscible displacement in porous media: Stability analysis of three-dimensional, axisymmetric disturbances with application to gravity driven wetting front instability, Water Resour. Res., 27(8), 1947-1956.
(9)   Parlange, J.-Y., 1975. On solving the flow equation in unsaturated soils by optimization: Horizontal infiltration, Soil Sci. Soc. Am. Proc., 39, 415-418.

Using Pore Size Distribution to Model the Unsaturated Hydraulic Conductivity of Soil

Bhabani S. Das and Gerard J. Kluitenberg. *Department of Agronomy, Kansas State University, Manhattan, KS 66506-5501, U.S.A.*

Introduction. Mualem (1) proposed a statistical approach for obtaining closed-form hydraulic conductivity functions. His approach resulted in the expression

$$K_r(\theta) = \frac{K(\theta)}{K_s} = S_e^l \left(\int_0^\theta d\theta/\psi \right)^2 \left(\int_0^{\theta_s} d\theta/\psi \right)^{-2} \qquad [1]$$

for the relative unsaturated hydraulic conductivity function where K_s is the saturated hydraulic conductivity, ψ is the soil water pressure head, and l is a parameter. The effective saturation S_e is defined $S_e = \theta/\theta_s = (\theta-\theta_r)/(\theta_s-\theta_r)$ where θ_r is the residual water content and θ_s is the water content at saturation. Equation [1] can be integrated if a water retention model (WRM) is available, but it has been implemented for only a restricted set of models that can be expressed explicitly as both $\psi(\theta)$ *and* $\theta(\psi)$. Mualem's approach has not been implemented for a number of popular retention models (2,3) that can be written explicitly only as $\theta(\psi)$. In this paper we show how information about the soil pore size distribution, $f(r)$, or the capillary pressure distribution $f(\psi)$ can be used to eliminate the need for an explicit $\psi(\theta)$ relationship. This permits the application of Mualem's approach to a new set of retention models.

Determination of $K(\psi)$ from $f(r)$ and $f(\psi)$. An alternative to [1] is an intermediate step that appears in the development of Mualem's theory:

$$K_r(\theta) = S_e^l \left(\int_{R_{min}}^R rf(r)\,dr \right)^2 \left(\int_{R_{min}}^{R_{max}} rf(r)\,dr \right)^{-2} \qquad [2]$$

Substitution of the capillary law ($\psi = C/r$) into [2] yields another alternative to [1]:

$$K_r(\psi) = S_e^l \left(\int_{-\infty}^\psi \frac{f(\psi)d\psi}{\psi} \right)^2 \left(\int_{-\infty}^{\psi_c} \frac{f(\psi)d\psi}{\psi} \right)^{-2} \qquad [3]$$

When a WRM is available, $f(\psi)$ can be obtained by differentiating $S_e(\psi)$ with respect to ψ (2,3). The capillary law can then be invoked to convert $f(\psi)$ to $f(r)$ if necessary. Alternatively, the capillary law can be used to convert $S_e(\psi)$ to $S_e(r)$ which yields $f(r)$ upon differentiation with respect to r. Note that the differentiation of $S_e(\psi)$ or $S_e(r)$ can often be completed for WRM's that only can be written explicitly as $\theta(\psi)$. Herein lies the advantage of the alternative forms [2] and [3]. Differentiation of a WRM can lead to $f(r)$ and $f(\psi)$ expressions that result in closed-form hydraulic conductivity functions upon substitution into [2] and [3]. We have tested this approach by deriving $f(r)$ and $f(\psi)$ for a popular WRM (4) that can be expressed explicitly both as $\psi(\theta)$ and $\theta(\psi)$. $K_r(\psi)$ was obtained by integration after substituting $f(r)$ into [2] and $f(\psi)$ into [3]. Both integrations yielded a result identical to that obtained by Mualem (1) when [1] was used.

$$S_e = \frac{1}{2}\text{erfc}\left[\frac{1}{\sigma\sqrt{2}}\left\{ \ln\left(\frac{\psi_c-\psi}{\psi_c-\psi_0} \right) - \sigma^2 \right\} \right], \; \psi<\psi_c ; \qquad S_e = 1, \; \psi \geq \psi_c \qquad [4]$$

New Hydraulic Conductivity Model. To further demonstrate the utility of this approach we present a new closed-form $K(\psi)$ expression that we derived from the three-parameter WRM that Kosugi (2) developed from a log-normal $f(\psi)$ relationship. Here ψ_c is the bubbling pressure, ψ_0 represents the mode of $f(\psi)$, and σ is a dimensionless parameter. The utility of this WRM has been established (2) but it is not possible to use the conventional approach of integrating [1] to obtain a closed-form expression for the hydraulic conductivity. Because of the complementary error function in [4], it is not possible to obtain the explicit expression for $\psi(\theta)$

31

needed in [1]. Using the approach outlined in the previous section, the $f(\psi)$ relationship corresponding to [4] was substituted into [3]. We have been successful in integrating [3] for the special case of $\psi_c = 0$ (note that [4] reduces to the WRM of (5) for the case of $\psi_c = 0$) which yields the closed-form expression

$$K_r(\psi) = S_e' \left[\frac{1}{2} \text{erfc} \left\{ \frac{1}{\sigma\sqrt{2}} \ln\left(\frac{\psi}{\psi_0} \right) \right\} \right]^2 \qquad [5]$$

Equation [5] represents a new form of hydraulic conductivity function based on Mualem's approach and is a new addition to the existing ones (viz., the power function type and the exponential type). It may further be observed that, in general, a power function type retention model yields a power function type conductivity model, an exponential type retention model yields an exponential type conductivity model, and a complementary error function type retention model yields an complementary error function type conductivity model. The need to evaluate the complementary error function in [5] makes this expression more complicated than other closed-form conductivity expressions. Nevertheless, efficient numerical schemes to evaluate the complementary error function are readily available, and it is important to make a variety of different soil hydraulic property models available. The ultimate usefulness of [5] will be determined by examining whether or not it can successfully predict measured unsaturated hydraulic conductivity relationships.

Summary. We have extended Mualem's method so that it can be used to obtain closed-form $K(\psi)$ expressions for WRM's that cannot be expressed explicitly as $\psi(\Theta)$. An example was presented in which our approach was used to derive a new closed-form conductivity expression corresponding to the WRM in [4] for the case $\psi_c = 0$. It may be possible to develop additional $K(\psi)$ relationships based on the developments presented herein.

Literature Cited.
(1) Mualem, Y. 1976. A new model for predicting the hydraulic conductivity of unsaturated porous media. Water Resour. Res. 12:513-522.
(2) Kosugi, K. 1994. Three-parameter lognormal distribution model for soil water retention. Water Resour. Res. 30:891-901.
(3) Brutsaert, W. 1966. Probability laws for pore-size distributions. Soil Sci. 101:85-92.
(4) Brooks, R. H. and A. T. Corey. 1964. Hydraulic properties of porous media. Hydrol. pap. 3, Civil Eng. Dep., Colo. State Univ., Fort Collins.
(5) Pachepsky, Ya. A., R. A. Shcherbakov, and L. P. Korsunskaya. 1995. Scaling of soil water retention using a fractal model. Soil Sci. 159:99-104.

Uneven Moisture Patterns in Sand, Loam, Clay and Peat Soils

Louis W. Dekker[1], Coen J. Ritsema [1], Tammo S. Steenhuis[2], and J.-Yves Parlange[2] . [1] *Winand Staring Centre for Integrated Land, Soil and Water Research, P.O. Box 125, 6700 AC, Wageningen, The Netherlands. [2]Cornell University, Department of Agricultural and Biological Engineering, Riley-Robb Hall, Ithaca, New York 14853-5701, U.S.A.*

Introduction. Various factors contribute to the spatial variaton of the soil water content and solute concentration in the vadose zone. Factors like vegetation, microtopography, water repellency and soil layering can cause irregular wetting and fingered flow (1,2,3,4). Accelerated transport of water and solutes towards the groundwater and surface water may be the result (5) and, therefore, it is essential to understand this phenomenon. We will present examples of soil moisture patterns due to vegetation, microtopography, water repellency, and soil layering.

Materials and Methods. During the last years sand, loam, clay, and peat soils in the Netherlands were sampled in vertical transects to study the spatial distribution of soil moisture content (2,6,7,8). Each transect was sampled at several depths with 25 to 100 steel cylinders (100 cm³), which were taken in close order. Per trench between 100 and 500 samples were collected. Each sample was used to determine soil water content, and its actual and potential water repellency (3). Measured soil water contents were used for constructing contour plots to visualize the soil water distribution within each seperate transect.

Results and Discussion. At the experimental station Aver-Heino near Zwolle, transects were sampled in a maize field during one growing season. Due to interception and stemflow, water was concentrated towards the roots. Also between the maize rows, higher soil water contents were found, caused by rainwater dripping to the ground from overhanging leaves. Microtopographical depressions further concentrated the dripped water. Clearly wetter soil areas are present near the roots and in between the maize rows (Fig. 1). The transport of water and solutes takes place mainly through these wetter portions. Redistribution of soil water from wet areas into dry areas is restricted because of the actual water repellency of the dry sand.

For a water repellent sandy soil with grass cover Ritsema and Dekker (4,7) presented evidence that water repellency greatly accelerates transport of water and solutes due to incomplete wetting of the soil. An example of the soil water content variation in horizontal direction at 5-10 cm depth is shown in Fig. 2. The peaks represent the wet fingers, and the valleys the dry soil in between. The horizontal line divides the actually wettable (above this line) and the actually water repellent soil parts. The ratio wettable and water repellent soil is time dependent.

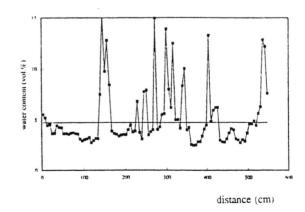

distance (cm)

Fig. 1. Influence of vegetation (maize) (micro)topography on the spatial distribution of water at the Aver-Heino experimental site

Fig. 2. Variation in water content at 5-10 cm depth and in a water repellent sandy soil with grass cover.

33

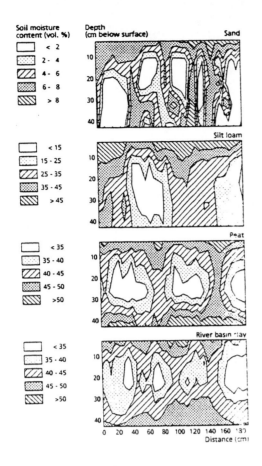

Soil moisture content (vol. %)

Sand:
- < 2
- 2 - 4
- 4 - 6
- 6 - 8
- > 8

Silt loam:
- < 15
- 15 - 25
- 25 - 35
- 35 - 45
- > 45

Peat:
- < 35
- 35 - 40
- 40 - 45
- 45 - 50
- >50

River basin clay:
- < 35
- 35 - 40
- 40 - 45
- 45 - 50
- >50

Depth (cm below surface)

Distance (cm)

Initially, scientists assumed fingered flow paths to occur in sandy soils only under specific conditions, such as water repellent sand or a poorly permeable sandy layer overlying a well permeable one. These flow paths, however, appear to occur in homogeneous, wettable sand as well. Fig. 3. shows the moisture distribution in a vertical plane of dune sand from the island of Terschelling. The wet top layer and the vertical flow paths are clearly visible. Between these paths, which have moisture contents of 6 to 10 vol.%, dry sand occurs with less than 2 vol.% of moisture. These large short-range differences in moisture content can only persist because sand shows a great degree of hysteresis. Not only did we find these fingered flow paths in wettable sandy soils, but also in water repellent silt loam, water repellent peat, and a slightly water repellent river basin clay soil (Fig. 3.). Especially in the silt loam, the wet paths are clearly wider than those in wettable and water repellent sand. Theories for predicting the widths of such wet paths indicate that a width of several centimeters is common in coarse sand, increasing up to 100 cm in finer-textured soils. The above-mentioned field measurements show that fingered flow patterns have a far more universal character than has been thought so far.

Acknowledgments. This research was carried out as part of Project C3-13: "Transport of Water and Solutes in Field Soils" of the Netherlands Integrated Soil Research Programme. Part of this work was carried out with support of the Environment Research Programme of the European Union, contract EV5V-0467. The European Union is thanked for providing the necessary travel funds.

Literature Cited.
(1) Dekker, L.W., and C.J. Ritsema. 1994. Fingered flow: The creator of sand columns in dune and beach sands. Earth. Surf. Process. Landforms, 19:153-164.
(2) Ritsema, C.J., and L.W. Dekker. 1994. Soil moisture and dry bulk density patterns in bare dune sands. J. Hydrol. 154:107-131.
(3) Dekker, L.W., and C.J. Ritsema. 1994. How water moves in a water repellent sandy soil. 1. Potential and actual water repellency. Water Resour. Res. 30:2507-2517.
(4) Ritsema, C.J., and L.W. Dekker. 1994. How water moves in a water repellent sandy soil. 2. Dynamics of fingered flow. Water Resour. Res. 30:2519-2531.
(5) Steenhuis, T.S., C.J. Ritsema, L.W. Dekker, and J.-Y. Parlange. 1994. Fast and early appearance of solutes in groundwater by rapid anf far-reaching flows. 15th World Congress of Soil Science, Acapulco, Mexico. July 10-16, 1994. Vol.2a:184-203.
(6) Dekker, L.W., and C.J. Ritsema. 1995. Fingerlike wetting patterns in two water-repellent loam soils. J. Environ. Qual. 24:324-333.
(7) Ritsema, C.J., and L.W. Dekker. 1995. Distribution flow: A general process in water repellent soils. Water Resour. Res. 31:1187-1200.
(8) Dekker, L.W., and C.J. Ritsema. 1995. Uneven moisture patterns in water repellent soils in the Netherlands. Geoderma (in press).

Estimation of Empirical Parameters for Soil-water Retention Curves

D. Dourado-Neto[1], D. R. Nielsen[2], J. W. Hopmans[2] and M. B. Parlange[2]. [1]*Fundação de Amparo à Pesquisa do Estado de São Paulo (Research Project number 92/5085-6). Department of Agriculture. ESALQ. University of São Paulo. C.P. 9. Piracicaba, SP. 13418-900. Brazil.* [2]*Hydrologic Science. Department of Land, Air and Water Resources. University of California. Davis, CA. 95616.*

Introduction. SWRC software was developed with the objective to estimate the empirical parameters of the soil-water retention curve, for different models, using least-squares method with the general iterative method of Newton-Raphson. The software allows for the following options: (i) estimation of initial values by anamorphosis and manual procedures, (ii) estimation of van Genuchten's model m for two different conditions: independent and dependent with the dependent case having restrictions described by Mualem's (m=1-1/n) model or Burdine's (m=1-2/n) model (12), and (iii) estimation of residual and saturated soil moisture by (a) a fixed value procedure (measured or any value fixed by the user), (b) by regression (with and without restriction), or (c) by an extrapolation method (8). This software can be useful for routine analysis of soil water retention data.

Material and Methods. SWRC software, version 1.00, was developed in VISUAL BASIC 3.0 for Windows environment (IBM or compatible) with thirteen models for soil-water retention curves (Table 1).

Table 1. Models for soil-water retention curves available in the SWRC software (version 1.00).

Model[1]	Author(s)	Model	Author(s)
$\theta = \theta_r + (\theta_s - \theta_r)e^{-\frac{\gamma\psi}{2}}\left(1 + \frac{\gamma\psi}{2}\right)^{\frac{2}{m+2}}$	Gardner (7)	$\psi = \alpha\left\{\exp\left[\beta(\theta - \phi)\right] - 1\right\}$	Simmons et al. (11)
$\theta = \theta_r + \dfrac{\theta_s - \theta_r}{(\alpha\psi)^\lambda}$	Brooks and Corey (1)	$\psi = \alpha\left\{\exp\left[\beta(\theta - \theta_s)\right] - 1\right\}$	Libardi et al. (9)
$\psi = \alpha(f - \theta)^\beta \theta^{-\omega}$	Visser (13)	$\theta = \theta_r + \dfrac{\theta_s - \theta_r}{\left[1 + (\alpha\psi)^n\right]^m}$	Van Genuchten (12)
$\theta = \theta_e + (\theta_{15} - \theta_e)\dfrac{\ln(\psi_e - \psi + 1)}{\ln(\psi_e - \psi_{15} + 1)}$	Rogowski (10)	$\theta = \theta_s \psi^{-\gamma\ln(\psi)}$	Driessen (5)
$\psi = \psi_e\left(\dfrac{\theta_s}{\theta}\right)^\lambda$	Campbell (3)	$\psi = -\alpha\exp(-\beta\theta)$	Exponential (2)
$\psi = \psi_{crit}\exp\left[\alpha(1 - Se)\right]$	Farrel and Larson (6)	$\psi = \alpha\theta^{-\beta}$	Power (2)

[1] $\psi = |\psi_m|$

The input file format is ASCII, where the first column refers to matric potential, and the following columns refer to soil water content.

The output permits an approximated r^2-value, as an indication of explained variance by the fitted model, the residual sum of squares to compare different models, and the analysis of variance for nonlinear regression. The F-test (4) is used to verify the selected model (Table 2). Estimated values of empirical parameters for each model, and graphs of the soil-water retention curve, relative hydraulic conductivity, water capacity and relative soil-water diffusivity are presented by the SWRC software.

Table 2. Analysis of variance (ANOVA).

Source of variation	Variation	Degrees of freedom	Variance	F-value	α^1
Explained by regression	$SSR = \sum_{i=1}^{k}\left(\hat{\theta}_i - \bar{\theta}\right)^2$	$V_1 = p - 1$	$MSR = \dfrac{SSR}{V_1}$	$F = \dfrac{SSR.V_2}{SSE.V_1}$	$\alpha = \int_F^\infty f(F).dF$
Unexplained (residual)	$SSE = \sum_{i=1}^{k}\left(\theta_i - \hat{\theta}_i\right)^2$	$V_2 = k - p$	$MSE = \dfrac{SSE}{V_2}$		
Total	$SST = \sum_{i=1}^{k}\left(\theta_i - \bar{\theta}\right)^2$	$k - 1$			

$$1 \; f(F) = \frac{\left(\frac{V_1}{V_2}\right)^{\frac{V_1}{2}}}{\beta(V_1, V_2)} F^{\frac{V_1}{2}-1}\left(1 + \frac{V_1}{V_2}F\right)^{-\frac{V_1+V_2}{2}}; \quad \beta(V_1, V_2) = \frac{\Gamma(V_1)\Gamma(V_2)}{\Gamma(V_1 + V_2)} = \int_0^1 x^{V_1-1}(1-x)^{V_2-1}dx$$

Literature Cited.

(1) Brooks, R.H. and A.T. Corey. 1964. Hydraulic properties of porous media. Colorado State University, Fort Collins, Hydrology Paper nr. 3.

(2) Bruce, R.R. and R.J. Luxmoore. 1986. Water retention: field methods. Methods of Soil Analysis, Part 1. Physical and Mineralogical Methods. 2nd edition. American Society of Agronomy and Soil Science of America. p. 663-686.

(3) Campbell, G.S. 1974. A simple method for determining unsaturated conductivity from moisture retention data. *Soil Science* 117: 311-4.

(4) Carnahan, B., H.A. Luther and J.O. Wilkes. 1969. Applied numerical methods. New York, John Wiley and Sons. 604p.

(5) Driessen, P.M. 1986. Land Use System Analysis. Wageningen.

(6) Farrel, D.A. and W.E. 1972. Larson. Modeling the pore structure of porous media. Water Resour. Res., 8:699-706.

(7) Gardner, W.R. 1958. Some steady state solutions of unsaturated moisture flow equations with application to evaporation from water table. Soil Sci., 85:228-232.

(8) Jong van Lier, Q. de and D. Dourado Neto. 1993. Valores extremos de umidade do solo referentes ao modelo de van Genuchten. R. bras. Ci. Solo, Campinas, 17:325-329.

(9) Libardi, P.L., K. Reichardt and V. F. Nascimento Filho. 1979. Análise da redistribuição da água visando a condutividade hidráulica do solo. *Energia Nuclear & Agricultura,* 1:108-22.

(10) Rogowski, A.S. 1971. Watershed physics: model of soil moisture characteristics. *Water Resources Research* 7: 1575-82.

(11) Simmons, C.S., D.R. Nielsen, D.R. and J.W. Biggar. 1979. Scaling of field-measured soil water properties. *Hilgardia* 47: 77-173.

(12) Van Genuchten, M.T. 1980. A closed-form equation for predicting the hydraulic conductivity of unsaturated soils. *Soil Science Society of America Journal* 44: 892-8.

(13) Visser, W.C. 1966. Progress in the knowledge about the effect of soil moisture content on plant production. Techn. Bull. 45. Institute for Land and Water Management Research, Wageningen, The Netherlands.

Applications of the CDE to Surface Accumulation and Leaching of Chemicals during Steady-State Flow in Variably Saturated Media

D.E. Elrick[1], M.I. Sheppard[2] and A. Nadler[3]. [1]*Department of Land Resource Science, University of Guelph, Guelph, Ontario, Canada N1G 2W1.* [2]*AECL Research, Whiteshell Laboratories, Pinawa, Manitoba, Canada R0E 1L0.* [3]*Institute of Soil and Water, ARO, Bet-Dagan, P.O. Box 6,50250, Israel.*

Introduction. The convective dispersion equation is often used as a functional description of solute transport in soils. The problem of salt or chemical accumulation near the soil surface during upward evaporative flux conditions for the initial condition where the salt or chemical substance was uniformly distributed with depth and the steady-state water content was a function of depth was examined previously (1). The analysis has been extended to the more general case where the initial distribution with depth is completely general, given for example as some function of depth (2). The more general case includes leaking underground storage tanks and other sources such as buried domestic or nuclear wastes where the source is below the soil surface. The rate at which finite amounts of salts or other chemicals either accumulate at the surface (net upward flux) or leach downwards (net downward flux) under steady-state flow conditions was examined. Expressions were derived for the spatial and temporal distribution of the resident and flux concentrations for buried Dirac delta function inputs as well as for more general distributions. As before (1), the expressions take into account linear adsorption and first-order decay or degradation of the chemical expanded to include the effects of a depth-dependent water content that is invariant with time. Equations for the equilibrium distribution near the surface are developed for finite amounts of contaminants initially present in the system.

A Four-Year Outdoor Soil Core Study on Capillary Rise of Radionuclides. A four-year study that monitored the leaching and capillary rise of Iodine (^{127}I) and several radionuclides has been reported by Sheppard and Hawkins (3). We focus on the capillary rise experiment. Additional details of the experiment are given in (4) and (5). Very briefly, intact soil cores ,0.051 m in diameter and on average 0.56 m long, were taken from a forested Brunisol (Dystrochrepts) soil in Manitoba, Canada. For the capillary rise (groundwater) cores, the elements were injected 0.1 m from the base, or about 0.46 m from the surface. The injection was carried out using a syringe that was inserted horizontally through a single small hole in the acrylic wall of the cores. The addition of ^{127}I and several radioactive elements could therefore be considered as a Dirac delta function or spike input at t = 0. The reservoirs of the acrylic cups of the groundwater cores were filled with simulated porewater and the water table was maintained at the base of the cores. The cores were wrapped, placed in a sleeve and buried in the ground in a fenced outdoor lysimeter field. The cores then received natural rainfall and experienced natural soil temperatures. After four years the cores were excavated, cut into 2 cm slices, the pore water was removed by centrifugation and then analyzed.The predicted resident concentration in the cores after the four-year experiment based on the relatively simple CDE solutions compare favourably with the data and gives a better prediction than the numerically intensive finite difference SCREM1 model. The model based on the variable water content with depth gives a better approximation at the lower depths where the concentrations are low (Fig. 1).

Deep Leaching of salts over a 30 year period. A field study on the effects of irrrigation on salt movement in a loessial soil in the semi-arid environment of N.W. Negev, Israel was reported by Margaritz and Nadler (6). We compare profiles at two sites, one which had been noncultivated and nonirrigated and the other which had been irrigated for 30 years. The two sites were several hundred metres apart but the assumption was made that the two profiles were identical. Samples were taken to a depth of about 22 m and we examined the chloride and volumetric water content profiles. The water content and chloride profiles at the uncultivated,unirrigated site were used to predict the chloride profile at the irrigated site. A downward flux of 4.3cm/yr gave a good prediction of the chloride concentrations profile in the irrigated site. The prediction based on the variable water content with depth gave a slightly better prediction of the chloride concentration in the irrigated profile than the one based on the mean water content (Fig. 2).

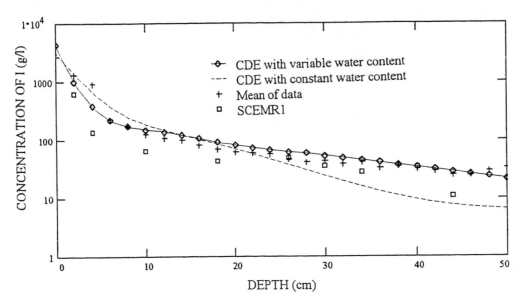

Fig. 1. Predictions of the I concentrations in a four-year outdoor soil core study on capillary rise compared to the data.

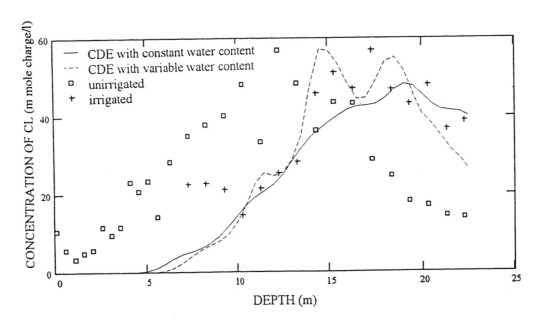

Fig. 2. Two predictions of the Cl concentration after 30 years of leaching in an irrigated deep profile based on a nearby unirrigated profile. The experimental data from the two profiles are shown.

Literature Cited.

(1) Elrick, D. E., A Mermoud,. and T. Monnier,. 1994. An analysis of solute accumulation during steady-state evaporation in an initially contaminated soil. J. Hydrol., 155:27-38.

(2) Elrick, D.E. M.I. Sheppard, A. Mermoud and T. Monnier. An analysis of surface accumulation of previously distributed chemicals during steady-state evaporation. Submitted to J. Hydrol.,1995.

(3) Sheppard, M.I., and J.L.,Hawkins,. 1991. A linear sorption / dynamic water flow model applied to the results of a four-year soil core srudy. Ecol. Modelling 55: 175-201.

(4) Sheppard, M.I., D.H. Thibault, and , J.H. Mitchell1,. 1987. Element leaching and capillary rise in sandy soil cores: Experimental results. J. Environ. Qual. 15: 273-284.

(5) Sheppard, M.I., and D.H Thibault., 1991., A four-year mobility study of selected trace elements and heavy metals. J. Environ. Qual. 20: 101-114.

(6) Magaritz, M and A. Nadler. 1993. Agrotechnically induced salinization in the unsaturated zone of loessial soils, N.W. Negev, Israel. Ground Water. 31:363-369.

Design and Analysis of Single-Core Experiments for the Determination of Unsaturated Hydraulic Properties

Boris Faybishenko, and Stefan Finsterle. *University of California, Department of Materials Science and Mineral Engineering, Earth Sciences Division, Lawrence Berkeley Laboratory, Berkeley, CA 94720*

Introduction. Modeling flow and solute transport in the unsaturated zone requires specifying values for unsaturated hydraulic conductivity (K) and water potential (P) as a function of water content (Q). A large number of laboratory and field methods have been developed over a number of years to determine unsaturated hydraulic properties. However, these methods are not only time-consuming and expensive, but they do not permit the determination of both curves during a single experiment. This inability to simultaneously determine both K and P as a function of Q often leads to parameter sets that are inconsistent and erroneous model predictions (1,5). The purpose of this paper is to present a new experimental procedure, which is especially designed for the determination of unsaturated hydraulic properties. Inverse modeling techniques have been incorporated in this approach to optimize design procedures and improve the analysis of laboratory data. The results are compared to semi-analytic solutions.

Methods. In this paper, we use results of laboratory experiments of undisturbed loamy soils performed on cores 22 cm in length and 15-18 cm in diameter (2,3). Single- and multi-step radial and axial flow experiments under both drying and wetting conditions were performed using different arrangements of the controlling ceramic cups and the monitoring tensiometers. A semi-analytic procedure was developed to calculate the unsaturated hydraulic conductivity based on flow measurements at the inlet or outlet ports. Water potential measurements on the wall of the core were also made. Using this approach, no assumptions had to be made in developing expressions of (P) and $K(P)$.

The transient flow of liquid and gas in the core has been studied by means of numerical simulations using the code TOUGH2 (6). The unsaturated hydraulic conductivity and water retention curves determined analytically have been compared to the values obtained using ITOUGH2 (4), which solves the inverse problem for multi-phase, multi-component flow systems. ITOUGH2 has also been used in a sensitivity analysis in order to optimize the instrumentation of the cores.

Results. Single-step and multi-step flow experiments under drying conditions yielded consistent water retention and relative permeability curves. Under wetting conditions, however, preferential flow induces air entrapment, and this causes hysteretic behavior in the unsaturated hydraulic properties. Measurements of flow rate and water potential were compared with those calculated using TOUGH2. The numerical simulations confirmed that the vertical component of the flow velocity can be neglected, and that the water pressure measured at the core wall (P_w) corresponds to the average water content (Q_{avg}) of the core. Therefore, these two values can be used in determining the water retention curve. A sensitivity analysis was performed using ITOUGH2 to evaluate the sensitivity of the Brooks-Corey parameters with respect to porosity, and the saturated hydraulic conductivity of both the soil and the ceramic cup. This analysis has shown that inverse modeling allows one to determine the unsaturated hydraulic parameters using only flow rate measurements across the controlling ceramic cup. Figure 1 shows an example of the comparison between experimental water retention data obtained from a single-step test and inverse modeling results.

Conclusions. A laboratory experiment has been designed for the simultaneous determination of unsaturated hydraulic conductivity and water potential as a function of liquid saturation. The experimental procedure is relatively simple and allows one to run concurrent tests in a timely fashion. Data from single-step and multi-step experiments were analyzed using a combination of analytical and inverse modeling techniques, and produced consistent results of high reliability. Furthermore, the proposed experimental design in combination with inverse modeling enables one to investigate the impact of temperature, water velocity, entrapped air, etc., on the hysteresis of (P) and $K(P)$ during several cycles of drying and wetting that are relevant to field applications.

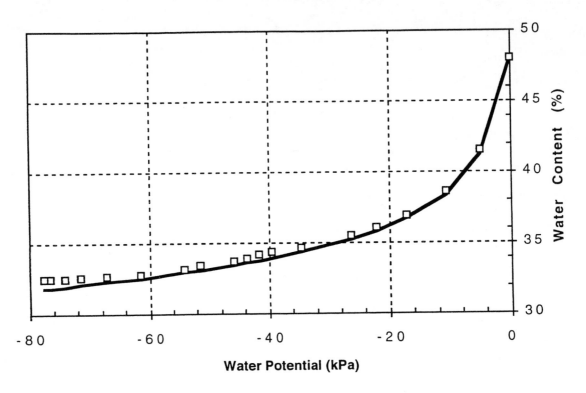

Fig.1. Example of comparison between data observed (symbols) in single-step experiment and calculated Brooks-Corey water retention curve obtained using inverse modeling.

Literature Cited.

(1) Clausnitzer, V., J.W. Hopmans, and D.R. Nielson, 1992, Simultaneous scaling of soil water retention and hydraulic conductivity curves, *Water Resour. Res., 28* (1), 19-31.
(2) Dzekunov, N.E., I.E. Zhernov, and B.A. Faybishenko, 1987, Thermodynamic methods of investigating the water regime in the vadose zone, Moscow, Nedra, 177.
(3) Faybishenko, B.A., 1986, Water-salt regime of soils under irrigation, Moscow,Agropromizdat, 314.
(4) Finsterle, S., 1993, ITOUGH2 User's Guide Version 2.2, Lawrence Berkeley Laboratory,report LBL-34581, Berkeley, California.
(5) Luckner, L., M.Th., van Genuchten, and D.R. Nielsen, 1989, A consistent set of parametric models for the two-phase flow of immiscible fluids in the subsurface, *Water Resour. Res., 25* (10), 2187-2193.
(6) Pruess, K., 1991, TOUGH2 - A general-purpose numerical simulator for multiphase fluid and heat flow, Lawrence Berkeley Laboratory, report LBL-29400, Berkeley, California.

Acknowledgment. This work was partially supported by the U.S. Department of Energy under Contract No. DE-AC03-76SF00098.

Monitoring Solute Removal from a Variably Saturated Soil During the Operation of a Pumping Well

P. Ferré[1], D.L. Rudolph[1], and R.G. Kachanoski[2]. [1]University of Waterloo, Waterloo, ON, Canada. [2]University of Guelph, Guelph, ON, Canada.

Introduction. The removal of contaminants from the vadose zone is an increasingly important task facing hydrogeologists. The operation of a pumping well to capture contaminated water in the saturated zone is common to many remedial designs. However, numerical analyses have shown that this approach is ineffective for complete removal of contaminant mass residing above the water table (4). The objective of this work was to monitor mass removal throughout the unsaturated and saturated zones at the field scale during the operation of a pumping well to test the effectiveness of the pump and treat approach.

Materials and Methods. A field site was selected in a relatively uniform sandy deposit on Canadian Forces Base Borden where the water table is approximately 2 meters below the ground surface. Bulk soil parameters were determined from previous investigations at the site and used in an existing numerical model (4) to design the spatial and temporal scales of the experiment and the monitoring program. The following steps were used to emplace a contaminant source in the vadose zone. Fresh water was infiltrated on the soil surface using a drip line irrigation system for 24 hours. This was followed by a two–hour pulse of saline (KCl) water and a final two–hour pulse of fresh water. After 220 hours of drainage, the surrogate contaminant source extended from approximately 140 cm below ground surface to below the water table. Then, a central pumping well was then operated for 250 hours at an extraction rate of 9 litres per minute.

Transient distributions of water and solute were monitored in three dimensions throughout the subsurface. A multilevel-TDR waveguide was designed to measure the water content profile (2). The multilevel-TDR waveguides and a DC–conductivity profiling system (3) were used to monitor the bulk electrical conductivity (EC) of the soil. Thermistors were used to account for the influence of temperature changes on the measured bulk EC. Together with the measured water content and temperature, the bulk EC was used to define the concentration of the electrolytic solute in the pore water, according to (1). Finally, suction lysimeters and piezometers were installed for direct comparison with the geophysical results.

Results and Discussion. Results from a measurement location within the salt-application area are shown in Figure 1. The water content, θ, as measured by TDR, was used in conjunction with the bulk EC, σ_b, measured with the DC electrical conductivity probe to determine the pore water conductivity, σ_s, following Archie's relationship (1):

$$\sigma_b = a\sigma_s\theta^2. \qquad [1]$$

The geometric factor for the DC probe, a, was determined by direct comparison with the EC of pore water samples collected with lysimeters and piezometers. In order to better represent the mass of contaminant present under varying water content conditions, the resulting solute concentration was multiplied by the soil water content to define the mass of solute present per unit volume of the porous medium. The calculated resident mass is plotted against depth and time and contoured in Figure 1. The water table elevation is also shown on Figure 1. Under drainage conditions, the capillary fringe at the site is approximately 30 cm thick and the soil reaches residual saturation for suctions above 145 cm of water.

The distribution of calculated resident mass (Fig. 1) shows the formation of the contaminant source straddling the water table with the majority of the solute mass originally residing within the capillary fringe and lower transition zone, defined as the region where the water content varies with elevation. Rapid removal of solute mass from below the water table is evident following the onset of pumping at approximately 240 hours. Solute mass appears to be continuously removed from the newly established capillary fringe and lower transition zone with continued pumping. However, some solute mass is clearly not removed by the pumping well. Figure 2 shows the total mass resident in the profile at each time calculated as the product of the average mass per unit volume of porous medium multiplied by the measurement interval. This resident mass reflects the solute mass in the measurement profile per unit area at the ground surface. This analysis shows that, for the conservative solute used in this experiment, approximately 10% of the solute mass was abandoned by the pumping well. The mass recovered and spatial distribution of solute mass within the subsurface following pumping are in excellent agreement with the results of numerical modeling based on the bulk soil parameters (4). Future analyses will focus on the influence of the regional flow on the movement of solutes above the water table.

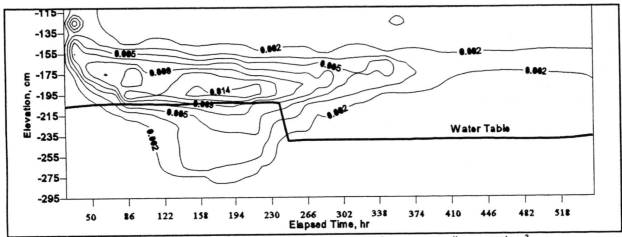

Fig. 1. Distribution of KCl mass per unit volume of porous medium, mg/cm³.

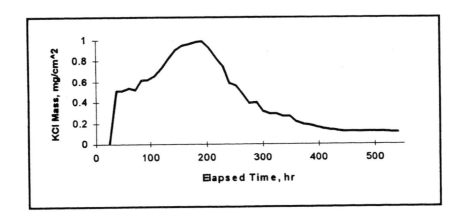

Fig. 2. Total KCl mass per unit ground surface area.

Literature cited.
(1) Archie, G.E. 1942. The electric resistivity log as an aid in determining some reservoir charac teristics. Trans. AIME 146:54-62.
(2) Ferré, P.A., D.L. Rudolph, and R.G. Kachanoski. 1994. A multilevel waveguide for profiling water content using time domain reflectometry. U.S. Bureau of Mines, SP 19-94:81-92.
(3) Schneider, G.W., S.M. DeRyck, and P.A. Ferré. 1993. Proceedings of SAGEEP, San Diego, CA.
(4) Xie, X., and R.W. Gillham. 1994. Modeling of dissolved-phase solute transport in the interface zone. Groundwater, (in submittal).

Petrologic Influences on the Unsaturated Hydraulic Properties of Geologic Materials

W.H. Frohlich[1], G.E. Fogg[1,2], and J. Mount[1]. [1] Department of Geology; University of California, Davis; CA 95616. [2] Hydrologic Science; University of California, Davis; CA 95616

Introduction. The geologically-based approach of Carle and Fogg (this volume) for characterizing spatial variability of the subsurface has a number of advantages. Applying this approach to vadose-zone characterization, however, requires that one relate vadose-zone properties, such as retention curve and hydraulic conductivity, to geologic attributes such as depositional facies (e.g., channel sands, overbank muddy sands, floodplain muds) either deterministically or stochastically. The purpose of this study is to observe quantitative and qualitative relationships, or lack thereof, between geologic attributes and vadose-zone properties at the pore and core scale so that the rich information on spatial variability that comes from geologic characterization can be used to greater advantage in vadose-zone studies.

The overall approach represented in this paper and that of Carle and Fogg (this volume) recognizes that much, if not most, of the vadose zone consists of geologic sediments laid down by fairly well-understood geologic processes often producing relatively structured material unlike the more perturbed mélange found in soils. Relative predictability of the geologic processes together with the sharp contrasts in sediment textures commonly produced by these processes leads us to the logical conclusion that geology is a necessary component of vadose-zone investigations if more than the upper meter is of interest. Unfortunately, the methods for rigorously applying geologic principles in vadose-zone hydrology are still in their infancy.

Materials and Methods. A total of 104 samples from several field site in the vicinity of Davis and Salinas, California were collected from the vadose zone. Using only outflow, and when necessary $h(\Theta)$, the saturated hydraulic conductivity (K_{sat}) and water retention curves were determined using the Multi-Step method of Eching and Hopmans [2]. Each core was examined macroscopic features (e.g. root holes, lamination, grading). We examined 38 cores petrographically. Examination of the arrangement of material at the pore scale was done using secondary electron, SEM imaging of Òintactó, stub-mounted samples, and optical microscopy and BSE* imaging of thin sections. Stub mounted samples and thin sections were prepared from both ends of each sample. Subsequent to sample preparation, the particle size distribution of all cores was determined using sieving and a laser granulometer. Gross mineralogy was determined using morphology and EDX** analysis [3].

Results. The sands grains of the Salinas sites were dominated by granitic minerals derived from the Gablin Mountains to the east. At depth samples contained significant amounts of chert grains from the Monterey Shale, deposited by the Salinas River and derived from the southern Salinas Valley. Sands of the Davis sites are quartzo-feldspathic of indistinct origin. In all samples the clays are predominantly smectitic. As expected, only the coarsest materials exhibited a strong link between texture and hydraulic properties. With increasing mean grain diameter a general trend of increasing K_{sat}, $d\Theta/dh$, and decreasing h was observed, with $d\Theta/dh$ and h evaluated at $\Theta=0.5$. However, as the amount of silt and particularly clay increases the scatter increases dramatically. Little to no correlation could be seen between the sorting or porosity and the hydraulic properties of the cores. This obviously shows that in sufficient quantity the fine grained materials, silt and clay, can dominate the unsaturated hydraulic properties.

Common in the petroleum literature are studies examining the influence of clay arrangement and mineralogy on permeability and mercury capillary pressure curves from samples with similar textures [4,5,6]. For comparative purposes the clay arrangements were categorized as pore lining, pore-throat bridging, pore-throat occluding, and pore filling (Fig. 1). Although, the total clay content and the relative size of the matrix material will play an important role, the arrangement of clays within samples of similar texture can be markedly different. It appears with increasing pore throat blockage, from pore lining, to bridging and then occlusion, there is a foreseeable decrease in K_{sat} and $d\Theta/dh$, and increase in h.

Discussion. If one only has data on grain size distribution, it appears that, at best one can do is to divide the media into coarse-grained (e.g., <~5% silt and clay), fine-grained (e.g., >~50% silt and clay), and intermediate textures, assign properties more or less deterministically to the coarse- and fine-grained endmembers, and perhaps stochastically to the intermediate textures to handle the greater uncertainty. Because the coarse- and fine-grained endmembers in geologic sediments commonly make up a significant volume fraction and have considerable spatial continuity, such a characterization of properties within a heterogeneity Òtemplate,Ó like those generated by Carle and Fogg (this volume), may be adequate.

The results showing a correspondence between clay arrangement in pores, K_{sat}, and the retention curve are encouraging, and warrant further study. The arrangement of clays within a given samples is a product of both deposition and the local post-depositional regime. This suggest that if one knows the origin of clays in a sample (e.g., depositional, illuvial, or authigenic), <u>such information can be</u> used to advantage in estimating vadose properties [7].

*BSE: backscattered electrons for SEM imaging of material densities in polished thin sections.

** EDX: energy dispersive X-ray, for use in a SEM to qualitatively identification of relative elemental fractions in minerals.

Acknowledgment. The authors would like to thank Norm Winter of the Geology Department for his irreplaceable assistance in the preparation of polished thin sections.

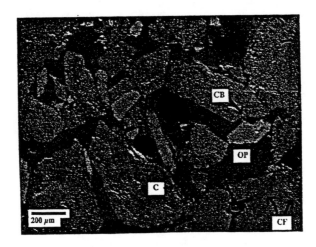

Figure 1. From a sample containing 2.5% clay, examples of clay bridging pore throats (CB), open pore throats (OP), clay cutan on grains (C), and clay filling pores between grains (CF). Note the clay bridge is broken and the cutan partally separated. This is frequently observed in thin sections containing swelling clay, drying being a necessarily step preceding resin impregnation [7].

Literature Cited.
(1) Carle, S.F. and Fogg, G.E. 1995. Markov approach to simulating geologic heterogeneity in alluvial sediments. (this volume)
(2) Eching, S.O. and Hopmans, J.W. 1993. Optimization of hydraulic functions from transient outflow and soil water pressure data. *Soil. Sci. Soc. Am. J.* 57(5): 1167-1175.
(3) Welton, J.E. 1984. SEM Petrology Atlas. *American Association of Petroleum Geologists.* Tulsa, Oklahoma. pp. 237
(4) Morris, K.A. and Shepperd, C.M. 1982. The role of clay minerals in influencing porosity and permeability characteristics in the Bridport sands of Wytch Farm, Dorset. *Clay Minerals* 17:41-54.
(5) Coskun, S.B., Wardlaw, N.C., and Haverslew, B. 1993. Effects of composition, texture and diagenesis on porosity, permeability and oil recovery in a sandstone reservoir. *J. Petr. Sci. Eng.* 8: 279-292
(6) Nelson, P.H. 1994. Permeability-porosity relationships in sedimentary rocks. *The Log Analyst* 35(3): 38-62.
(7) Moraes, M.A.S. and De Ros, L.F. 1992. Depositional, infiltrated and authigenic clays in fluvial sandstones of the Jurassic Sergi Formation, Reconcavo Basin, northeastern Brazil In [ed.s Houseknecht, D.W. and Pittamn, E.D.] Origin, Diagenesis, and Petrophysics of Clay Minerals in Sandstone. *SEPM Special Pub. 47.* pp. 197-208.

Preferential Flow During Solute Transport Through Naturally Heterogeneous Soil Columns

Hesham M. Gaber, Heiko W. Langner, Jon M. Wraith, and William P. Inskeep. *Plant, Soil & Environmental Sciences Department, Montana State University, Bozeman, MT 59717*

Introduction. Preferential flow in heterogeneous soils may result in more rapid leaching of pollutants through soils than would be predicted using transport models based on the local equilibrium assumption (LEA)(1). The potential for preferential flow in a specific soil is determined by the pore size distribution and the presence and continuity of macropores (2). However, significant water flow through macropores will only occur when they are water-filled (2,3). The water status of macropores is closely related to soil matric potential (ψ_m); consequently, there is a relationship between ψ_m and the occurrence of preferential flow. Knowledge of this relationship could help to identify and predict soil water conditions under which accelerated contaminant transport is likely. Preferential flow can be described using a two-domain (mobile vs. immobile water) model and is synonymous with transport-related nonequilibrium (TNE) (4,5). The bicontinuum model of the convective-dispersive equation (CDE) can be used to determine the degree of TNE for nonsorbing solutes. The primary objective of the current study was to determine relationships between soil water conditions (i.e., matric potential, pore water velocity) and the degree of TNE in naturally structured soils.

Materials and Methods. Two series of miscible displacement experiments were conducted using intact cores (26-cm length, 15-cm ID) of Amsterdam silt loam soil (Fine-silty, mixed Typic Haploboroll). The first series was performed at soil-water (Darcy) flux levels of 0.04, 0.27, and 0.86 cm h^{-1}. A solution pulse of $\approx$ 0.12 pore volumes containing ^{3}H$_2$O-ring labeled ^{14}C-atrazine in 3 mM CaCl$_2$ was applied to each column using a precision syringe pump. Column effluent was collected over time using a fraction collector. ^{14}C and ^{3}H activities were determined in a 1 mL effluent. The second series was conducted using a modified disk permeameter (5) to impose constant potential conditions (0 to 25 cm water tension) within the columns. A solution pulse of $\approx$ 0.7 pore volumes containing ^{3}H$_2$O-PFBA in 3 mM CaCl$_2$ was applied to each column through the disk permeameter at desired ψ_m. Eluent fractions were analyzed for ^{3}H$_2$O (scintillation) and PFBA (ion chromatography). Solute breakthrough curves (BTCs) were fitted to the CDE using CXTFIT (6). The degree of TNE in each experiment was evaluated by comparing model fits using LEA and the two-domain nonequilibrium models.

Results and Discussion. Two series of miscible displacement experiments were conducted using large intact soil cores. The first series was performed using a ^{3}H$_2$O-atrazine pulse at different pore water velocities. The asymmetrical shape and the left-handed displacement of the ^{3}H$_2$O BTCs indicate the existence of TNE under fast and medium pore water velocities (2.16 and 0.74 cm h^{-1} respectively). The asymmetrical shape and increased tailing of atrazine BTCs was influenced primarily by sorption-related nonequlibrium (SNE) at slow pore water velocity (0.12 cm h^{-1}) and a combination of both TNE and SNE at medium and fast pore water velocities. The second series was conducted over a range of soil matric potentials using a modified disk permeameter and a ^{3}H$_2$O-PFBA pulse. A transition from equilibrium to nonequilibrium conditions was observed between ψ_m of -5 to -10 cm. These potentials correspond to maximal effective pore radii of 0.15 to 0.3 mm. Comparison among experiments for a single intact core shows increasing TNE with increasing pore water velocities. However, we observed a range in degree of TNE among different soil cores exhibiting similar pore water velocities. A probable explanation is variability in pore size distribution among the cores. TNE was consistently observed at $\psi_m \geq$ -5 cm, independent of intact soil core replicate and pore water velocity within the range of 1 to 2 cm h^{-1}.

Ψ_m (cm)	PWV (cm/h)	Soil Core	β ω	Shape of ³H₂O Breakthrough Curve (0 to 4 PV)
ponded	1.00	PN11	0.2 0.518	
-2	1.14	PN12	3.7E-4 13.0	
-3	0.96	PN13	0.43 3.3	
-5	1.24	PN13	0.74 23.4	
-10	1.06	PN12	1 4.4E+5	

Fig. 1. Matric potential influence on ³H₂O breakthrough curves for experiments having similar pore water velocities

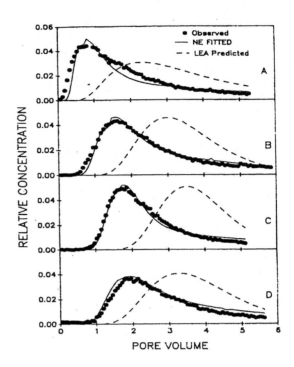

Fig. 2. Observed atrazine breakthrough curves (BTCs), fitted BTCs using the nonequilibrium (NE) model, and predicted BTCs using the local equilibrium assumption (LEA): (A) fast, (B) medium (volumetric soil water content = 0.44), (C) medium, and (D) slow pore water velocity

Literature Cited.

(1) Gaber, H.M., W.P. Inskeep, S.D. Comfort, and J.M. Wraith. 1995. Nonequilibrium transport of atrazine through large intact soil cores. Soil Sci. Soc. Am. J. 59:60–67.

(2) Seyfried, M.S., and P.S.C. Rao. 1987. Solute transport in undisturbed columns of an aggregated tropical soil: preferential flow effects. Soil Sci. Soc. Am. J. 51:143–1443.

(3) Brusseau, M.L., and P.S.C. Rao. 1990. Modeling solute transport in structured soils: A review. Geoderma. 46:169–192.

(4) van Genuchten, M.Th., and R.J. Wagenet. 1989. Two-site/two-region models for pesticide transport and degradation: Theoretical developments and analytical solution. Soil Sci. Soc. Am. J. 53:1303–1310.

(5) Langner, H.W., H.M. Gaber, J.M. Wraith, W.P. Inskeep, and B. Huwe. 1994. Preferential flow through intact soil columns: Effects of matric potential. P. 243. *In* Agronomy Abstracts. ASA, Madison, WI.

(6) Parker, J.C., and M.Th. van Genuchten. 1984. Determining transport parameters from laboratory and field tracer experiments. Virginia Agric. Exp. Stn. Bull. No. 84-3.

Kinematic Approach to Structural Impacts on Flow in Highly Saturated Porous Media

P. Germann, Soil Science Section, Department of Geography, University of Bern, Hallerstrasse 12, 3012 Bern, Switzerland.

Introduction. The genial simplicity of the approaches by Darcy and Richards to flow in porous media is based on the ignorance of soil structure. Consequently, these two approaches are limited when soil structure affects flow. Flow in soils with and without structure can be approached by kinematic wave theory (1).

Theory. The following set of equations apply to highly saturated soils. The balance equation of a kinematic wave is:

$$q / \partial t + c \, \partial q / \partial z = 0 \qquad\qquad [1]$$

where the volume flux density is $q = b \cdot w^a$. The parameters are: b m s^{-1} is conductance, w m^{3}m^{-3} is soil moisture participating in the flow process, and a is a dimensionless exponent; the celerity c m s^{-1} is defined as

$$c = dq/dw = a \, b^{1/a} \, q^{(a-1)/a} = a \, b \, w^{(a-1)} \qquad\qquad [2]$$

A wetting front is released at $t = 0$ under the initial conditions of $w(z,0) = q(z,0) = 0$ for $0 \leq z \leq \infty$ and the upper boundary conditions of $t \leq 0$: $q(0,t) = 0$, $0 \leq t \leq t_s$: $q(0,t) = q_s$, and $t \geq t_s$: $q(0,t) = 0$ (input of a rectangular pulse). The celerity of the wetting front is

$$c_w = q / w = dz_w / dt = b^{1/a} \, q^{(a-1)/a} = b \, w^{(a-1)} \qquad\qquad [3]$$

The wetting front arrives at depth Z at:

$$t_w(Z) = Z / (b^{1/a} q_s^{(a-1)/a}) = Z / (b \, w^{(a-1)}) \qquad\qquad [4]$$

A draining front is released at $t = t_s$. It arrives at Z at:

$$t_D(Z) = t_s + Z / (a \, b^{1/a} \, q_s^{(a-1)/a}) = t_s + Z / (a \, b \, w^{(a-1)}) \qquad\qquad [5]$$

$q(Z,t)$ and $w(Z,t)$ become:

$0 < t \leq t_w(Z)$:	$q(Z,t) = 0$;	[6a]		$w(Z,t) = 0$		[6b]
$t_w(Z) \leq t \leq t_D(Z)$:	$q(Z,t) = q_s$;	[7a]		$w(Z,t) = (q_s/b)^{1/a}$		[7b]
$t \geq t_D(Z)$:	$w(Z,t) = w(Z,t_D) \, [(t_D(Z)-t_s) / (t-t_s)]^{1/(a-1)}$					[8]

The exponent **a** can be estimated with Eq. [8] from the decrease of soil moisture with time after the passing of the draining front at Z, $t \geq t_D(Z)$.

The exponent **a** is proposed as a fractal dimension of porous media flow: $a < 2$: turbulent flow; $a = 2$: flow in cylindrical tubes; $2 < a < 3$: intermediate behavior; $a = 3$: flow along planar fissures and cracks; $3 < a < \approx 10$: flow shifts increasingly towards dispersive behavior; $a > \approx 10$: flow is dominated by the dispersive properties of the porous medium (i.e., as expressed in the Richards-Equations). The variation of **a** is due to the structure of the porous medium **and** the upper boundary conditions: In gerneral, the higher q_s the smaller **a** (2).

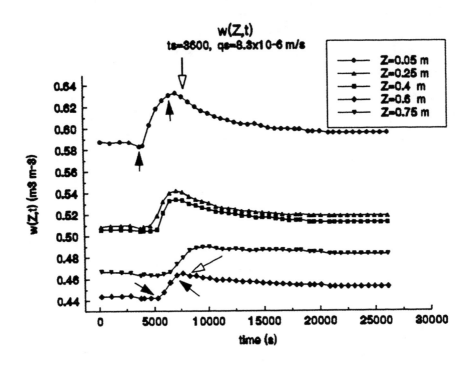

Figure 1: $w(Z,t)$. Solid arrows: t_{w1} and t_{w2};
open arrows: t_D

Experiment, Results and Discussion . A sprinkling experiment (application rate 30 mm/h, duration 1 h) was performed in situ on a block of soil (1x 1m). $w(Z,t)$ at depths Z of 5, 25, 40, 60 and 75 cm was measured with a TEKTRONIX TDR probe *(3)* (see Fig. 1). The application of Eq. [8] to the falling limbs of $w(Z,t)$ yielded $29 \le$ **a** ≤ 99 and $6.8 \le$ **b** $\le 2\times10^{26}$ms^{-1}, indicating dispersive flow (see Tab.1).

The application of Eq. [4] to the arrivals of the wetting fronts at depths Z,

$$a=1+\ln[Z/(t_w b)]/\ln[w], \quad [9]$$

using $6.8 \le b \le 2\times10^{26}$ ms^{-1} yielded the range of $3.9 \le$ **a** ≤ 20. The lower part of this range indicates preferential flow.

Table 1: Results of the data analysis

Depth m	t_{w1}[1] s	t_{w2}[1] s	t_D s	a —	b m/s	r^2 —
0.05	4068	6480	6983	29.5	6.76	0.96
0.25	4680	6480	7020	38.5	7.0×10^5	0.95
0.40	5292	6480	7088	39.9	4.6×10^6	0.98
0.60	5292	8868	7776	60.6	9.2×10^{15}	0.99
0.75	6480	9468	10620	98.9	2.0×10^{26}	0.81

[1] t_{w1} is the first observed increase of soil moisture,
        t_{w2} is the time when the supposed plateau was arrived

Literature Cited.

(1) Germann, P.F. 1990. Preferential flow and the generation of runoff. 1. Boundary- layer flow theory. Water Resources Research, 26, 3055- 3063.

(2) DiPietro, L., and LaFolie, F. 1991. Water flow characterization and test of a kinematic-wave model for macropore flow in a highly contrasted and irregular double-porosity medium. Journal of Soil Science , 42, 551-563.

(3) Bürgi, Th. 1993. Bestimmung der Dynamik des Bodenwassers mittels Tensiometern und TDR-Sonden unter Feldbedingungen. M.Sc.-Thesis, Faculty of Science, University of Bern (Switzerland), 91 p.

Calculation of Temperature Effects on Contact Angles at Porous Solids and Their Capillary Pressure Functions

Steven A. Grant, *U.S. Army Cold Regions Research and Engineering Laboratory, 72 Lyme Road, Hanover, NH 03755-1290 USA.*

Introduction. Numerous studies[4][6] have found that the Philip–de-Vries model [7] underestimates temperature sensitivities of experimentally determined capillary-pressure-functions (CPFs). This fact, coupled with reported observations of non-zero contact angles of liquids in porous media[1,5] prompted me to explore the notion that temperature sensitivities of wetting coefficients contribute to the temperature effect on CPFs. I assumed that the capillary pressure, p_c (Pa), was related to the effective radius, r (m), at a particular degree of saturation, s_1 (1), of the porous medium by the Kelvin equation:

$$p_c = p^g - p^l = 2\gamma^{lg} \cos\theta \, / \, r(s_1) \tag{1}$$

where p^g and p^l are the pressures in the gas (non-wetting) and liquid (wetting) phases, respectively (Pa); γ^{lg}, the liquid-gas interfacial tension (N m^{-1}); and θ, contact angle (rad). At a given s_1, the derivative of p_c with respect to temperature, T (K), is:

$$\left(\frac{\delta p_c(s_1)}{\delta T} \right)_{s_1} = \frac{p_c(s_1)}{\gamma^{lg}} \frac{\delta \gamma^{lg}}{\delta T} + \frac{p_c(s_1)}{\cos\theta} \frac{\delta \cos\theta}{\delta T} \tag{2}$$

The derivative $\delta \cos\theta / \delta T$ can be estimated via the relation of Harkins and Jura[2], yielding the following simple relation for the capillary pressure divided by its temperature derivative:

$$\frac{p_c(s_1)}{\left(\dfrac{\delta p_c(s_1)}{\delta T} \right)_{s_1}} = - \frac{\Delta_g^l h_s}{C_1} + T \tag{3}$$

where

$$\Delta_g^l h^s$$

is the enthalpy of immersion per unit area (J m^{-2}) and C_1, a constant, the value of which must be determined from the value of κ at a reference temperature, T_r. Integration of a rearranged equation (3) yields the following relation for p_c as function of observational temperature, T_t:

$$p_{c_{(t-T_t)}} = p_{c_{(t-T_r)}} \frac{-\Delta_g^l h^s / C_1 + T_t}{-\Delta_g^l h^s / C_1 + T_r} \tag{4}$$

Materials and Methods. I analyzed the data of Salehzadeh[8], who measured the imbibition and drainage water-retention curves of two soils (a Plano silt loam and an Elkmound sandy loam) and a glass-bead sample at four temperatures between 5°C and 40°C. For selected degrees of saturation, cubic-spline interpolants were fitted to 4 p_c-T pairs.[3] The values of $(\delta p_c(s_i)/\delta T)_{s_i}$ and $p_c(s_i)$ were calculated from the fitted cubic-spline parameters. For each imbibition or drainage event, all the s_i-T-p_c data were fitted globally by non-linear regression[9] to the van Genuchten's equation[10] modified by equation (4).

Results and Discussion. Plots of the

$$s_1 - T - p_c(s_i)\left(\frac{\delta p_c(s_i)}{\delta T}\right)_{s_i} \tag{6}$$

surfaces were strikingly planar. This planarity reflected the generally linear trend of

$$p_c(s_i)\left(\frac{\delta p_c(s_i)}{\delta T}\right)_{s_i} \tag{7}$$

with temperature. The slopes of these surfaces was estimated by linear regression and found to be ≈ 1, as predicted by equation (3). Van Genuchten's equation modified by equation (4) could be fitted precisely by non-linear regression to the s_1-T-p_c data. The temperature sensitivities of these CPFs appear to be due largely to capillarity. Temperature-sensitive contact angles appear to explain most of the temperature effects on these CPFs.

Literature Cited.
(1) Demlehner, U. 1991. The contact angle of liquids in porous media in Characterization of Porous Solids 2. Edited by F. Rodriguez-Reinoso, J. Rouquerol, K.S.W. Sing, and K.K. Unger. pp. 97-104. Elsevier, Amsterdam, The Netherlands.
(2) Harkins, W.D. and G. Jura. 1944. Surfaces of solids, 12., An absolute method for the determination of the area of a finely divided crystalline solid. J. Am. Chem. Soc. 66:1362-1366.
(3) IMSL, Inc. 1989. IMSL Math/Library User's Manual. IMSL, Inc., Sugar Land, Texas.
(4) Jury, W. and E. Miller. 1974. Measurement of the transport coefficients for coupled flow of heat and moisture in a medium sand. Soil Sci. Soc. Am. Proc. 38:551-557.
(5) Lu T-X., J.W. Biggar, and D.R. Nielsen. 1994. Water movement in glass bead porous media, 1, Experiments of capillary rise and hysteresis. Water Resour. Res. 30:3275-3281.
(6) Nimmo, J.R. and E.E. Miller. 1986. The temperature dependence of isothermal moisture vs. potential characteristics of soils. Soil Sci. Soc. Am. J. 50:1105-1113.
(7) Philip, J.R. and D.A. de Vries. 1957. Moisture movement in porous materials under temperature gradients. Trans. Amer. Geophys. Union. 38:222-232.
(8) Salehzadeh, A. 1990. The temperature dependence of soil-moisture characteristics of agricultural soils. 191 pp. Ph.D. dissertation, Univ. of Wisconsin, Madison. Dissertation Abstracts No. AAD91-01559.
(9) SAS Institute, Inc. 1985. SAS user's guide, Statistics, Version 5 edition. SAS Institute, Inc., Cary, North Carolina.
(10) van Genuchten, M.Th. 1980. A closed-form equation for predicting the hydraulic conductivity of unsaturated soil. Soil Sci. Soc. Am. J. 4:892-898.

Combined Effects of Landscape Topography and Moisture-Dependent Anisotropy on Soil-Water Flow

T.R. Green[1], D.L. Freyberg[2], J.E. Constantz[3]. [1]CSIRO Division of Water Resources, Private Bag, Wembley, WA 6014, Australia. [2]Stanford University, Dept of Civil Engineering, Hydrology and Environmental Fluid Mechanics, Stanford, CA 94305, USA. [3]USGS, Water Resources Division, Menlo Park, CA 94025, USA.

Introduction. Spatial and temporal variability in soil-water movement affects soil development, soil-water availability to plants, and the transport and fate of near-surface contaminants. Small-scale soil layers create anisotropy in the upscaled hydraulic conductivity. Previous work has demonstrated the potential magnitude of moisture-dependent anisotropy values (1, 2) and the implications for vertical migration and spreading of point-source contamination (3). Additionally, numerical simulations have shown that moisture-dependent anisotropy caused lateral unsaturated flow consistent with tracer measurements beneath a planar slope (4). In the present study, we emphasize the effects of topographic curvature on vertical unsaturated flow. We address the questions: How does the combination of undulating topography and moisture-dependent anisotropy affect soil-water flow and the distribution of recharge rates in both space and time? Will topographic focusing cause the long-term, net recharge beneath undulating terrain to be greater than predictions assuming vertical flow only?

Methods. Two-dimensional unsaturated flow was numerically simulated in cross sections beneath hypothetical undulating landforms. A finite-element model (5) was modified to incorporate moisture-dependent anisotropy functions. Each finite element contained many small-scale (local) soil layers that were homogenized (upscaled) over the element. The Mualem-van Genuchten model (6) was used to represent the local hydraulic conductivity and water retention. Upscaled soil hydraulic properties were derived to approximate spatial averages for known sequences of local soil layers. We also modified Yeh et al.'s (1) anisotropy function for application to relatively dry soils (2). Landscape topography determined the dip in the principal directions of conductivity for each element. For spatially uniform infiltration, the combination of this dip (i.e., slope) and anisotropy caused lateral unsaturated flow. Steady state flow was simulated beneath several undulating landforms with less than 5 meters of relief and a drainage boundary condition at a depth of 10 meters. Transient simulations used twenty years of meteorologic data from Ender's Lake, Nebraska, USA . Spatially uniform rainfall infiltration was specified in the absence of overland flow. Thus, all lateral flow occured in the subsurface.

Results and Discussion. Topographic curvature caused soil-water accumulation beneath swales and depletion beneath crests. Figure 1 shows patterns of saturation and Darcy flux vectors beneath one landform during steady flow. Zones of soil-water accumulation acted as preferential recharge areas. Different combinations of landscape topography, anisotropy, and steady recharge rates determined the spatial distributions of both Darcy flux and upscaled soil-water contents. At steady state, Darcy flux varied by as much as four orders of magnitude from a minimum flux beneath a topographic crest to a maximum flux beneath a swale or bench. Vertical flux at a depth of 2 meters was strongly correlated ($R^2 = 0.96$) with topographic curvature for relatively complex landforms. For the transient experiments, temporal variations were greatest in zones of topographically focused recharge. Results from our bare-soil simulations did not support the hypothesis that landscape topography would increase the net recharge over time and space. Soil hydraulic properties, rather than landscape topography, affected the net recharge.

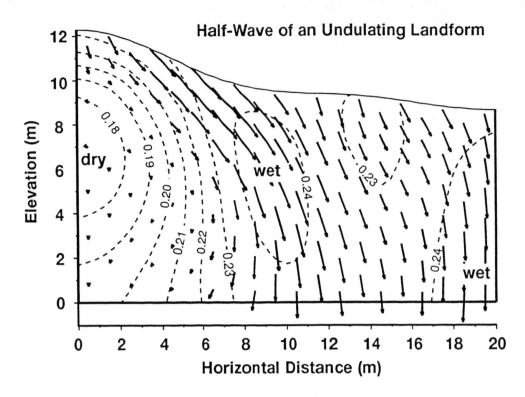

Figure 1. Example of Darcy flux vectors and volumetric water content contours for steady flow.

Conclusions. The simulated unsaturated flow processes caused topographically focused vertical flux in the absence of overland flow or perched groundwater. Spatial and temporal variability in flux beneath undulating terrain adds complexity to the problem of estimating net recharge. Simulated flow patterns have important ramifications for leaching of pesticides from the root zone, solute transport to groundwater, and pedologic processes related to observed, but previously unexplained spatial variability in soil development. Future work should include feedback mechanisms between unsaturated flow, soil development, plant growth, and evapotranspiration.

Literature Cited.
(1)   Yeh. T.-C.J., L.W. Gelhar, and A.L. Gutjahr. 1985. Stochastic analysis of unsaturated flow in heterogeneous soils, 2, Statistically anisotropic media with variable a. *Water Resour. Res.* 21(4):457-464.
(2)   Green, T.R. and D.L. Freyberg. In press. State-dependent anisotropy: Comparisons of quasi-analytical solutions with stochastic results for steady gravity drainage. *Water Resour. Res.*
(3)   Polmann, D.J., D. McLaughlin, S. Luis, L.W. Gelhar, and R. Ababou. Stochastic modeling of large-scale flow in heterogeneous unsaturated soils. *Water Resour. Res.* 27(7):1447-1458.
(4)   McCord, J.T., D.B. Stephens, and J.L. Wilson. 1991. Hysteresis and state-dependent anisotropy in modeling unsaturated hillslope hydrologic processes. *Water Resour. Res.* 27(7):1501-1518.
(5)   Yeh, G.T. 1987. A three-dimensional finite element model of water flow through saturated-unsaturated media. Oak Ridge National Laboratory, Rept. No. ORNL-6386.
(6)   van Genuchten, M.Th. 1980. A closed-form equation for predicting the hydraulic conductivity of unsaturated soils. *Soil Sci. Amer. J.* 44(5):892-898.

Delineation of the Time Domains for "Early-Time" and "Steady-State" Analyses of Pressure Infiltrometer Data

P.H. Groenevelt[1], D.E. Elrick[1], and B.P. Odell[1]. [1]*Department of Land Resource Science, University of Guelph, Guelph, Ontario, Canada N1G-2W1.*

Introduction. The determination of the field-saturated hydraulic conductivity, K_{fs}, using a Guelph Pressure Infiltrometer, typically relies on the establishment of steady state flow conditions. For media with very low values of K_{fs}, such as the compacted clay liners of landfill sites, the establishment of steady state flow conditions may take months. It then becomes opportune to rely on "early-time" observations. The time domain in which this early-time data analysis is possible can be delineated from the domain in which steady-state analysis is appropriate.

Theory. Elrick and co-researchers have developed equations for the determination of the field saturated hydraulic conductivity, K_{fs}, using the Guelph pressure infiltrometer for both steady state infiltration ("long-time") and sorption dominated ("early-time") processes (1,2).

By equating these expressions for the discharge rate into the soil surface, Q, one obtains a value for the time which is both, the longest possible "early-time", and the shortest possible "long-time" for the analysis of K_{fs}. The equation for that time value reads:

$$t = \frac{S_H^2}{4}\left[\left(\frac{H}{\pi r G} + \frac{1}{2}\right)K_{fs} + \frac{\phi_m}{\pi r G}\right]^{-2} \qquad [1]$$

where ϕ_m is the matric flux potential, and S_H, the sorptivity at the ponded head H, is given by (3):

$$S_H^2 = 2\Delta\theta K_{fs} H + \frac{\Delta\theta\phi_m}{b} \qquad [2]$$

Results and Discussion. We present here an impression of the values of the longest theoretical early-time readings (t), that makes K_{fs} determination possible in that domain, in relation to two dominant field and clay liner conditions, i.e. K_{fs} and the change in volumetric water content between initial and field saturated conditions ($\Delta\theta$). In Figure 1 this is shown for the following simplifying conditions: H (ponded head of water) = 0; G (geometric shape factor (2)) =0.5; b (factor adapted from (3)) = 0.5.

With $S_H^2 = S_o^2 = 2\phi_m\Delta\theta$ equation (1) can then be written as:

$$t = W\Delta\frac{\theta}{K_{fs}} \qquad [3]$$

$$W = 2\left(\frac{\pi r\sqrt{\alpha^*}}{\pi r\alpha^* + 4}\right)^2 \qquad [4]$$

whereand $\alpha^* \equiv K_{fs} / \phi_m$.

In Figure 1, log time has been plotted as a function of pK_{fs} ($pK_{fs} = -\log_{10}K_{fs}$) for different values of $\Delta\theta$ taking log W = -2. Figure 1 may aid in the sometimes arduous task of determining the hydraulic conductivity in the field. For wet, compacted clay liners, with $K_{fs} < 10^{-9}$ ms^{-1} and $\Delta\theta > 0.01$ the time domain for "early-time" analysis is approximately the first day of measurement.

It is now of practical interest to investigate what the effect is on the boundary of the time domains if one relaxes the condition H = 0 (m). Whilst keeping G = 0.5 and b = 0.5 equation (1) now approximately reduces to:

$$t = \frac{\Delta\theta}{K_{fs}}\left\{\frac{\pi^2 r^2}{8(H + 1/\alpha^*)}\right\} \qquad [5]$$

With r (radius of the infiltrometer ring) = 0.05 m and $\alpha^* = 4.5$ (i.e. log W ≅ -2, see Figure (1)), the effect of changing H (m) is shown in Figure (2) for $\Delta\theta = 0.01$.

Literature cited.

(1) Elrick, D.E., W.D. Reynolds. 1992. Infiltration from Constant-Head Well Permeameters and Infiltrometers. Soil Sci. Soc. Am. Special Publication no.30 1-24.

(2) Elrick, D.E., G.W. Parkin, W.D. Reynolds, and D.J. Fallow. 1995. Analysis of Early-Time and Steady-State Single Ring Infiltration Under Falling Head Conditions.Water Resour.Res. (In press).

(3) White, I., M.J. Sully. 1987. Macroscopic and Microscopic Capillary Length and Time Scales From Field Infiltration. Water Resources Research. 23(8):1514-1522.

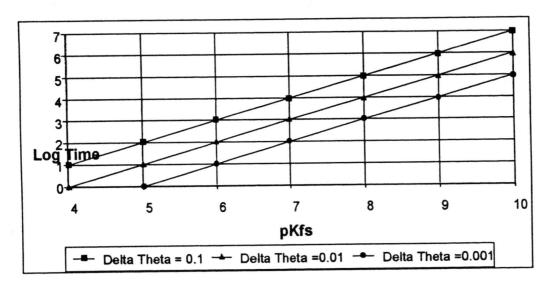

Fig. 1. pK$_{fs}$ versus log time for varying delta theta.

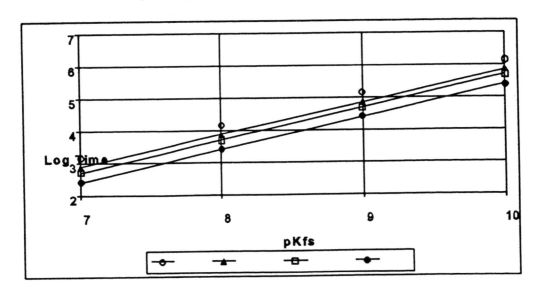

Fig. 2. pK$_{fs}$ versus log time for varying applied head with constant $\Delta\theta = 0.01$.

Recent Advances in the Evaluation of Water Supply Capability of Soil to a Single Root and the Measurement of Soil Moisture Content through the Measurement of Volumetric Heat Capacity with a Simple Probe

Michihiro Hara, *Iwate University, Morioka, Japan 020*

INTRODUCTION. This paper discloses our recent advances in two subjects stated in the title both related to soil moisture regimes. The first topic is a mathematical method for evaluating the water supply capability of soil when water is supplied to a single root. The effects of non-linearity in the soil's hydraulic properties according to the moisture change were taken into account. Previous works(2) on this evaluation assumed the linearity of soil properties against moisture change. However, so enormous moisture change might have occurred in the adjacent soil when water was absorbed strongly with a fine root that non-linearity of soil's hydraulic properties became no more negligible. Conditions that allowed the use of Boltzmann's transformation were assumed in this research. Thus, constant rate absorption of water through a straight line was assumed to be occurred in infinite homogeneous soil under gravity negligible conditions. Unique exponential relationships between water diffusivity and soil moisture content were assumed. The objectives of the study were to provide more compact mathematical expressions that could be used in calculating the water supply capability of soil accurately than usual numerical integration. The second topic of this work is a newly developed simple thermal-probe that measures volumetric heat capacity of materials without taking samples to laboratory. Because volumetric heat capacity of soil is linearly related with moisture content in it, this new probe may be useful in measuring soil moisture content periodically both field and laboratory conditions. Previous thermal-probe(1) for soil moisture detection measured not the heat capacity but the thermal conductivity. Since thermal conductivity of soil related nonlinearly with the change of soil moisture content precise calibrations were required for individual soil. In developing the new thermal probe, homogeneous and infinite soil conditions were assumed. The objectives of the study were to provide design conditions of the probe type heat capacity sensor to measure soil moisture content with desired accuracy.

MATERIALS, METHODS, RESULTS, DISCUSSION AND CONCLUSION.

Water Supply capability of soil

The water supply capability of soil is defined by Hara as the ratio of reached amount of water in a determined duration of time to the depression of matric water potential at the root surface when the plant root absorbs water in a determined chronological pattern. For evaluating this, absorption of water by a straight line in a homogeneous and infinite soil under gravity negligible conditions is considered. A cylindrical coordinate r is taken assuming symmetric absorption with respect to z-axis and uniform absorption along the z-axis. Unique exponential relationships in hydraulic properties were assumed in this study. Governing equations for flow rate and water content, conditions and relating functions are as follows.

A variable y for the Boltzmann's transformation and new variables u and Q lead a set of ordinary differential equations.

Successive approximations starting with known solution for constant diffusivity were compared with usual numerical integration and found to give good accuracy after two steps of approximations.

Probe Type Sensor to Measure Volumetric Heat Capacity

If amount of heat Q_m is released instantaneously from point source (m=3) at origin, temperature f_m at radius r and elapsed time t will become

This formula is integrated to obtain temperature for arbitrary shape of heat source and time course. Especially for instantaneous line and plane sources, m is taken as 2 and 1, respectively. The temperature f_m has a peak $f_m.peak$ at $t=t_{m.peak}$. Volumetric heat capacity C_0, thermal diffusivity D_0 thus thermal conductivity K_0, too, could be obtained from these relationships.

Effects of finite duration of heating time as well as heater radius and length were studied numerically for the case of line source, where K_0 is free from detecting position since m=2. A pair of parallel cylindrical surface sources revealed practically useful because this arrangement of heaters reduced the sensitivity of small error in detecting positions. Detection of fm.peak is less sensitive to possible measuring errors than that of tm.peak, thus the volumetric heat capacity is most suitably detected with this probe among the 3 thermal properties. The larger radius of heaters reduces the maximum heater temperature, thus reduces undesirable soil water movement, too, with small measurement error.

LITERATURE CITED.

(1) DeVries,D.A. and Peck,A.J.1958. On the cylindrical probe method of measuring thermal conductivity with special reference to soils.(I) Aust.J.Phys.11:255-271.
(2) Gardner,W.R.1960. Dynamic aspects of water availability to plants. Soil Sci.89:63-73.

Conditional Stochastic Analysis of Solute Transport in Soils

Th. Harter[1], D. Zhang[2]. [1] *Department of Land, Air, and Water Resources, University of California -Davis, Parlier, CA 93648.* [2] *Daniel B. Stephens and Associates, Inc., 6020 Academy NE, Suite 100, Albuquerque, NM 87109*

Introduction. Solute transport in soils is subject to uncertainty due to a variety of practical problems: Accuracy of measurements, correct interpretation of data, and most of all - spatial variability of soil properties. Stochastic methods have been used to account for uncertainty about soil properties when predicting solute transport. In situ field measurements locally reduce the uncertainty about soil properties. It is hypothesized that such measurements therefore also lead to reduced uncertainty about solute transport. To account for specific local information about soil properties, stochastic properties are conditioned. Two methods are here used to achieve such conditioning: Monte Carlo simulation (1) and Eulerian-Lagrangian analysis (2) modified for unsaturated transport (3). We briefly introduce both methods and apply them to hypothetical field sites. The examples are used to contrast the two methods and to illustrate the effect of saturated hydraulic conductivity and soil water tension measurements on solute transport uncertainty. We consider vertical, two-dimensional steady state unsaturated flow and transient transport in a Gardner-type soil (4) with a local solute dispersivity of 4 cm longitudinally and 1 cm laterally.

Approach 1: Monte Carlo simulation (MCS). In MCS the steady-state flow and transient transport equations are solved numerically (5) many times over a fine grid, each time with a different set of random K_s (saturated hydraulic conductivity) and α (pore-size distribution) parameter fields. These fields are conditioned on either K_s or Ψ (soil water tension) data or both. For conditioning, first order covariance and cross-covariance functions are used that were derived from spectral analysis. Computational efficiency is achieved by applying a combination of spectral and finite element methods (1). The need for a fine grid arises from the necessity to accurately resolve the spatial distribution of Ψ, which may be highly nonlinear. We determined that 300 realizations gave accurate mean concentration distributions $<c(x,t)>$, even in very heterogeneous soils and with few conditional data.

Approach 2: Eulerian-Lagrangian approximation (ELA). The Eulerian-Lagrangian transport model avoids Monte Carlo simulations by solving directly for $<c(x,t)>$. ELA is based on the assumptions that unconditional mean flow is uniform and conditional velocity fluctuations are mild. Spatial variability of porosity is assumed to be negligible. Initially, a mass of solute is injected uniformly over an area A_o. The resulting mean concentration at early time is obtained analytically based on the conditional probability distribution function of the velocity. After the average solute plume has traveled approximately ½ correlation scales, a pseudo-Fickian approximation for the mean concentration is used. In the resulting transport equation, the conditional dispersion tensor is given, to first-order in the conditional variance of velocity, by

$$D(x,t) \approx \int_0^t <v'(x)\ v'^T(<\chi(\tau)>)>\ d\tau \qquad [1]$$

where $<v'v'^T>$ is the conditional velocity covariance and $<\chi(\tau)>$ is the conditional mean position at time τ of a particle, known to be at a downstream location x at later time t. The velocity moments $<v>$ and $<v'v'^T>$, conditioned on K_s, α, and/or Ψ data, are obtained via cokriging. Appropriate first order approximations of unconditional velocity covariance and unconditional cross-covariances between hydraulic conductivity parameters, head, and velocity are derived spectrally (1). Once the conditional velocity moments are available, the integral in [1] is evaluated numerically and the pseudo-Fickian transport equation is solved by the finite element method (2). It is solved on a grid that is generally coarser than that required for Monte Carlo simulations. This is so because $<v>$ and $<c>$ are much smoother than v and c, respectively, and therefore require lesser resolution. The more data are available per finite area or volume, the less smooth $<v>$ and $<c>$

become, and grid resolution must then be made finer. Much of the computational effort is expended on the evaluation of **D** in [1].

Comparison Procedures. The two methods are applied to various hypothetical soils with varying degrees of anisotropy and heterogeneity. Correlation scales of K_s and α are chosen to be identical and vary from 100 to 300 cm horizontally and from 50 to 100 cm vertically. The variance of unsaturated hydraulic conductivity varies from 0.25 to 1. Within each soil type, three "field sites" are chosen from which data are collected. Each "field site" is a single random realization (K_s, α, and corresponding finite element solution of Ψ) of the same soil type. Either K_s or Ψ data or both are "measured" in a dense grid with measurement points less than one correlation scale apart in both horizontal and vertical direction. With such high density of conditioning data, mean concentration plumes are significantly distorted, particularly in the more heterogeneous soils. Differences between the results from MCS and ELA are therefore more pronounced. In actual applications, data density for conditioning will likely be less. The output time for which we compared results is $t = 3\lambda_z/\langle v_z \rangle$, where λ_z is the vertical correlation scale and $\langle v_z \rangle$ is the mean vertical flux. Average soil water tension was relatively low with mean vertical flux of 0.15-0.25 cm/minute.

Results and Discussion. It was shown that strong, nonlinear dependence of unsaturated hydraulic conductivity on soil water potential introduces complex interdependencies between different types of measurement data with respect to their ability to reduce flow and transport prediction uncertainty (1). Stronger conditioning yields less symmetric, less uniform $\langle c(x,t) \rangle$. Both methods are able to capture these relationships. For example, both methods yield less uniform $\langle c(x,t) \rangle$ when conditioning with K_s data than when conditioning with Ψ data, if soils are relatively wet. Between MCS and ELA, overall spatial distribution of $\langle c(x,t) \rangle$ is similar for the dominant features. Depending on the particular data, MCS and ELA will both yield a mean plume moving slow or very fast in its initial stage, or moving diagonally to the right or left from the source. Smaller scale features of the mean plume tend to be ignored by ELA when compared to MCS. In general we find that MCS mean plumes lag behind the ELA plume due to differences between the first order analytical and the nonlinear MCS unconditional mean velocity. The MCS plumes show greater lateral spread, less longitudinal dispersion, and greater distortion due to conditioning. Differences are attributed to the unknown small-scale effects of numerical dispersion in MCS on one hand, and to the first order approximation in ELA on the other hand. High data sampling density dictated that finite element discretization in ELA is as fine as in MCS. The differences in CPU time between a single MCS and a single ELA implementation were more than an order of magnitude.

Conclusion. Eulerian-Lagrangian approximation provides a useful tool for conditional simulation of variably saturated flow and transport, particularly in soils with moderately heterogeneous flow conditions. In particular with respect to computational demand, this method is preferable to Monte Carlo simulation. Monte Carlo simulation, on the other hand, provides much higher flexibility. This is particularly important with respect to the choice of boundary conditions and parametric soil properties.

Literature Cited.
(1) Harter, Th., 1994, Unconditional and Conditional Simulation of Flow and Transport in Heterogeneous, Variably Saturated Porous Media, Ph.D. dissertation, University of Arizona.
(2) Zhang, D., S.P. Neuman, 1994, Eulerian-Lagrangian analysis of transport conditioned on hydraulic data: 1. Analytical-numerical approach, Water Resour. Res. 31:39-51.
(3) Harter, Th., D. Zhang, 1995, Conditional Prediction of Transporting Unsaturated, Heterogeneous Porous Media: Monte Carlo Simulation versus Eulerian-Lagrangian Theory, Proceedings of the XXI General Assembly, IUGG, July2-14, 1995.
(4) Gardner, W. R., 1958, Some steady state solutions of unsaturated moisture flow equations with applications to evaporation from a water table, Soil Sci., 85, 228-232.
(5) Yeh, T.-C. J., R. Srivastava, A. Guzman, Th. Harter, 1993, A numerical model for two-dimensional water flow and chemical transport, Ground Water, 32:2-11.

Tensiometer Slug Tests: A New Method to Measure In-situ Hydraulic Conductivity

Masaki Hayashi[1], Garth van der Kamp[2] and Dave Rudolph[1]. [1]*Waterloo Centre for Groundwater Research, Univ. of Waterloo, Waterloo, Ontario, N2L 3G1, Canada.* [2]*National Hydrology Research Institute, Saskatoon, Saskatchewan, S7N 3H5, Canada.*

Introduction. Hydraulic conductivity, K, becomes a function of pressure head, ψ, under unsaturated conditions. A number of methods are available to measure $K(\psi)$ in the field (1), however, many have limitations particularly in clayey soils due to very slow drainage. A new method involves the application of an instantaneous pressure change (slug) to the head space of a tensiometer and subsequent monitoring of the pressure decay. This provides an alternative method in low-permeability settings where most existing methods are not applicable.

Theory. $K(\psi)$ is estimated by matching the solution of Richards equation to measured pressure decay. The numerical solution of the Richards equation shows that the dependence of K on ψ during a slug test is negligible for a small change of ψ, e.g. 0.1-0.5 m, suggesting that the application of linear slug test theory to tensiometers is reasonable for small perturbations. A classic Hvorslev (2) type method is used in calculation of K, in which $\psi(t)$ is modeled as:

$$[\psi(t) - \psi_E] / [\psi_0 - \psi_E] = \exp(-KFt) \qquad [1]$$

where ψ_E is the equilibrium pressure head, ψ_0 is the initial pressure head, and F is a shape factor determined by the geometry of a porous cup and the gage sensitivity. The equation [1] gives an estimate of K for ψ_E. The soil water storage is neglected in [1], however, the numerical solution shows that the degree of errors due to storage is within that of experimental errors.

Methods. Slug tests were performed in a 20L plastic bucket filled with soil collected from the St.Denis Research Site near Saskatoon, Saskatchewan. The saturated hydraulic conductivity, K_S, of the soil measured by the constant head test was 1×10^{-7} m/s. Two tensiometers were installed, one with a stainless steel porous cup (outer diam.=6.4mm, $K_S = 9 \times 10^{-7}$ m/s) and another with a standard ceramic porous cup (o.d.=23mm, $K_S = 2 \times 10^{-8}$ m/s). Each tensiometer was sealed on top by a rubber stopper. A pressure transducer was connected to each through a syringe needle and valve (Fig.1). The soil was initially saturated and was then progressively dried to induce changes in ψ. The soil moisture release characteristic curve was obtained by weighing the bucket at several stages during the drying process, and the data were fit to the van Genuchten (3) equation. A slug was introduced by connecting a known volume of air at known pressure (usually atmospheric) to the head space, and decay was monitored. In some cases a pair of slug tests were conducted by introducing positive and negative pressure changes in the head space.

Results and discussion. Fig.2 shows the results of a pair of slug tests at $\psi_E = -1.6$m. Circles and crosses represent decays of the positive and negative slugs, respectively, while the solid curve is the best fit of [1]. Fig.2 confirms that the dependence of K on ψ is negligible. The values of conductivity measured by the ceramic porous cup tensiometer at various values of ψ are summarized in Fig.3, in which the solid curve represents the values predicted by the van Genuchten (3) equation and the dashed line represents the conductivity of the ceramic cup. The measured conductivity follows the trend of the predicted conductivity reasonably well provided the soil conductivity is well below the conductivity of the ceramic cup. As the tensiometer measures the local conductivity around the porous cup, the method is potentially susceptible to some degree of error due to the skin of foreign material introduced during the installation. It was found in the experiment that the stainless steel cup was particularly susceptible to such bias and unreasonably low values of K were estimated. Extra care must be taken in the installation of tensiometers to be used for slug tests.

Conclusion. This new method gives a reasonable estimate of $K(\psi)$ in the wide range of ψ, provided that the skin effect is not serious. Typically it only takes about an hour to make a measurement, and only a fraction of mL of water is injected to the soil, which makes the method attractive for continuous monitoring of the hydraulic conductivity.

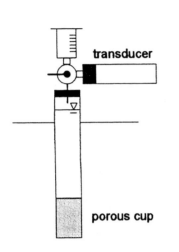

Fig.1 Tensiometer set up.

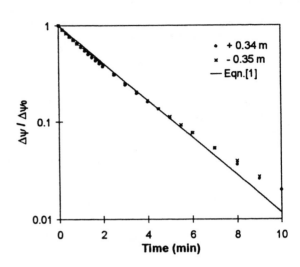

Fig.2 A pair of tests with positive (+0.35m) and negative (-0.34m) slug.

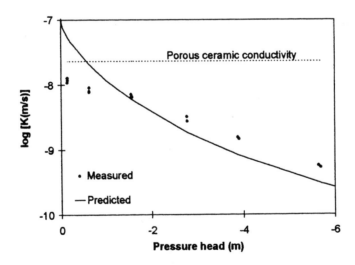

Fig.3 Relationship between pressure head and conductivity.

Literature Cited.
(1)   Green, R.E., Ahuja, L.R. and Chong, S.K. 1986. Hydraulic conductivity, diffusivity, and sorptivity of unsaturated soils: Field methods. In Methods of Soil Analysis, Part I. Physical and Mineralogical Methods - Agronomy Monograph no.9., 2nd ed., Soil Sci. Soc. Am. 771-798.
(2)   Hvorslev, M.J. 1951. Time lag and soil permeability in ground-water observations. Corps of Engineers, U.S. Army, Waterways Experiment Station, Bull. No.36. 50pp.
(3)   van Genuchten, M.Th., 1980. A closed-form equation for predicting the hydraulic conductivity of unsaturated soils. Soil Sci. Soc. Am. J. 44:892-898.

Field-Scale Transport of Water and Bromide in a Cracking Clay Soil

R.F.A. Hendriks[1], W. Hamminga[1], K. Oostindie[1] and J.J.B. Bronswijk[2]. [1]DLO Winand Staring Centre for Integrated Land, Soil and Water Research, Wageningen, Netherlands. [2]National Institute of Public Health and Environmental Protection, Bilthoven, Netherlands.

Introduction. One of the main factors that hampers the modeling of field-scale solute transport is the occurrence of preferential flow (1). Preferential flow is the process in which a part of the applied water and solutes flows rapidly to the subsoil and groundwater, bypassing a large volume of unsaturated soil. It can be caused by such varying phenomena as macropores, unstable wetting fronts, topography, crops, tillage, and it can occur in a wide variety of soils ranging from sand to clay soils (2,3). In heavy clay soils, preferential flow is mainly caused by soil structure. From reviews (1,4) it becomes apparent that experiments on heavy clay soils to study solute transport at the field scale are lacking. The objective of this study is to present field-scale water and solute transport in a heavy clay soil under dry conditions, where preferential flow through structural voids is expected to be important (5). Therefore a bromide tracer has been applied to a grassland field on a cracking heavy clay soil.

Materials and Methods. The experimental field was situated in the riverine area of the Netherlands. The land use was pasture for cattle grazing. The field was tile drained (0.95 m depth, 15 m spacing). The soil can be classified as a very fine clayey, mixed illitic-montmorillonitic, mesic, Typic Fluvaquent (6). The clay content ranges from 51 to 60%. From earlier experiments, these soils are known to exhibit distinct swelling and shrinking (7,8). Bromide was applied to the field by spraying 0.05 mm of water with a concentration of 360 g L^{-1} KBr. Thus bromide application equalled 12.0 g m^{-2}. The bromide was applied homogeneously under dry conditions when shrinkage cracks were visible present. At 0, 6, 46, 209, 335 and 572 days after bromide application 15 soil columns (1 m lenght, 0.10 m diameter) were sampled. The columns were devided into 20 samples of equal length. The volumetric water content, dry bulk density and bromide content of each sample were determined. Water samples were weekly taken from groundwater tubes at 1.2-1.4 m below the soil surface. Drain outflow was sampled automatically at regular time intervals. Groundwater and drain water samples were analyzed for bromide content.

Results and Discussion. The bromide tracer was applied under dry conditions (Fig. 1A). Following the bromide application, during the autumn period (days 0-46) the soil was gradually wetted. During the winter period (day 46-209) water contents further increased. After the summer (day 335) the soil had dried out again. The second winter again resulted in wet conditions (day 572). The field-average solute transport velocity was very high. After 6 days the average solute peak had already reached a depth of 0.55 m with only 34 mm of net precipitation (Fig. 1B). Subsequently, bromide advanced further in the soil. The average recovery of bromide in the top meter of the soil subsequently decreased from 95% at the start 104%, 84%, 70%, 41%, and 25% after 6, 46, 209, 335, and 572 days, respectively, as a result of downward bromide transport. A low volumetric water content at the moment of tracer application with large shrinkage cracks caused rapid preferential flow of water and bromide through the unsaturated zone. This resulted in maximum bromide concentrations in groundwater and drain discharge directly after tracer application. The encountered bromide concentrations decreased during this field study, but precipitation events were followed by a rise in the average bromide concentration in groundwater and drain discharge. On the basis of the results we can consider three flow domains in the studied heavy clay soil (Fig. 2). First, the macropores such as large continuous shrinkage cracks, with rapid transport of water and solutes through the unsaturated zone. This transport amounted only a few percent, but resulted in maximum bromide concentrations in groundwater and drain discharge. Second, the smaller more tortuous mesopores with lateral and vertical transport through the soil between the large macropores. This was quantitatively the most important transport. It is characterized by high flow velocities, high spatial variability, significant lateral transport, and low mobile water volumes (6%). Third, the micropores in the soil aggregates, that play only an indirect role in transport, i.e., via solute retardation as a result of convection and diffusion into the aggregates. Slow transport via these small pores can declare that after 1.5 year still 25% of the applied bromide remained in the sampled zone. Computer simulation models to describe the solute transport mechanisms in heavy clay soils in a realistic way, should take these three domains into account.

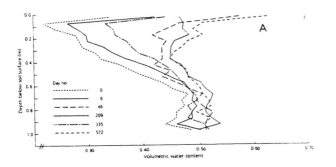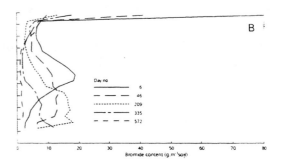

Fig. 1. Measured field-average water (A) and bromide content profiles (B). Each profile represents the average of 15 soil columns. Day 0 is September 26, 1991.

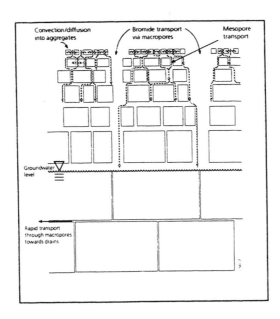

Fig. 2. Water and solute transport mechanism in a structured heavy clay soil.

Literature Cited.
(1) Jury, W.A. and H. Flühler. 1992. Transport of chemicals through soil: Mechanisms, models, and field applications. Adv. Agron. 47:141-201.
(2) Beven, K. and P. Germann. 1982. Macropores and water flow in soils. Water Resour. Res. 18:1311-1325.
(3) Gee, G.W., T. Kincaid, R.J. Lenhard, and C.S. Simmons. 1991. Recent studies of flow and transport in the vadose zone. U.S. Natl. Rep. Int. Union Geod. Geophys. 1987-1990 Rev. Geophys. 29:227-239.
(4) Beven, K.J., D.E. Henderson, and A.D. Reeves. 1993. Dispersion parameters for undisturbed partially saturated soils. J. Hydrol. 143:19-43.
(5) Bronswijk J.J.B., W. Hamminga, and K. Oostindie. 1995. Field-scale solute transport in a heavy clay soil. Water Resour. Res. 31:517-526.
(6) Soil Survey Staff, Soil Taxonomy. 1975. A basic system for soil classification for making and interpreting soil surveys. Agric. Handb. U.S. Dep. Agric. Sci. Edu. Adm. 436, 754 pp.
(7) Bouma J., L.W. Dekker, and C.J. Muilwijk. 1981. A field method for measuring short-circuiting in clay soils. J. Hydrol. 52:347-354.
(8) Bronswijk J.J.B. 1988. Modeling of water-balance, cracking and subsidence of clay soils. J. Hydrol. 97:199-212.

NMR Relaxation of Water Saturated Porous Matrices: Quantification of Microporosity

Z.R. Hinedi, A.C. Chang and M. A. Anderson. *Dept. Soil and Environmental Sciences, University of California, Riverside, CA 92521*

Introduction: Nuclear Magnetic Resonance (NMR) provides a non-invasive probe of the molecular organization of water in heterogeneous porous media. In a porous matrix the relaxation rate of water is enhanced with respect to that of bulk water due to van der Waals interactions effective over near neighbor distances λ (3 Å) at the water-grain interface (1, and references therein). The NMR mesurement consists of exciting the spin system in the water and recording its relaxation back to equilibrium over the delay time (t) which can vary between 10 μs to more than 500 ms. The measured relaxation rate is a linear combination of the bulk relaxation rate ($1/T_b$) and the relaxation rate at the surface ($1/T_s$) weighted by surface-to-volume ratio (s/v) which holds the pore size information. In a porous matrix (e.g. porous silica, soils.) which consists of a collection of pores where fluid relaxes at different rates, the contribution of all the relaxation rates ($1/T_i$) to the magnetization decay A(t) is weighed by the volume (w_i) of fluid that relaxes at each rate:

$$A(t) = \sum_i \exp(-t/T_i) \,/\, \sum_i w_i \qquad [1]$$

where

$$\frac{1}{T_i} = \frac{1}{T_b} + \frac{\lambda}{T_s}\frac{m}{l}$$

In its continuum form equation (1) can be written as follows

$$A(t) = \int_0^\infty P(l)\,\exp\left[-t\left(\frac{1}{T_b} + \frac{\lambda}{T_s}\frac{m}{l}\right)\right] \qquad [2]$$

where
P(l) is defined as the probability distribution that a molecule of liquid is in a pore with the size range $l + dl$ and
l is the pore size
T_b is the relaxation of the bulk water
T_s is the relaxation at the interface
The volume and the surface of a pore are proportional to l^3 and l^2 respectively and
s/v = m/l where m depends on the geometry of the pore.

Materials and Methods: Model porous silica (Davisil) as well as soils representative of different textural and mineralogical compositions in addition to an aquifer material were used in this study. Their characteristics have been described elsewhere (1). Porous beds were packed into 5 mm NMR tubes, evacuated then saturated under a CO_2 atmosphere. The longitudinal relaxation data were collected on a Joel FX200 NMR. In the inversion recovery pulse sequence (180°-t-90°), the delay time t was varied between 10 ms and 5000 ms depending on the porous matrix. The contribution of the bulk relaxation rate $1/T_b$ (measured for bulk water in the absence of any packed porous matrix) is subtracted from A(t) prior to solving the magnetization data in equations [1 and 2]. Nonlinear regression was used to solve equation [1]. The computer code CONTIN (2) was used to obtained a regularized relaxation distribution from equation [2].

Results and Discussion. Nuclear relaxation measurements of water in model silica beds with known pore sizes revealed two relaxation times: the larger one is attributable to interparticle water while the smaller one which yielded pore sizes comparable to those reported in the porous Davisil silica product literature is attributed to intraparticle pores (Figure 1). The NMR relaxation measurements of the water saturated soils and aquifer material collected under saturated conditions allowed the detection of bi-and multimodal pore size distributions which were in general agreement with the mineralogical and textural compositon. The average pore domains ranged in size from 12 nm to 3000 nm (Figure 2). The pore sizes obtained by NMR in model silica materials and in the Borden aquifer agreed with pore sizes reported in the litererature using N_2 adsorption methods.

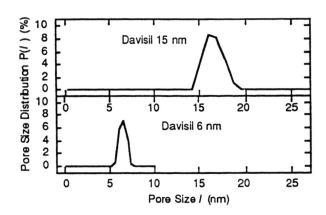

Fig. 1 Intraparticle porosity in Davisil silica beads.

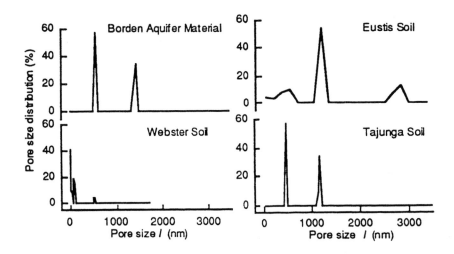

Fig. 2. Pore size distribution in soils and aquifier materials (published in part in Hinedi et al 1993).

Literature Cited.
(1) Hinedi, Z.R., Kabala, Z.J., Skaggs, T.H., Borchard, D.B., Lee, R.W.K. and Chang, A.C. 1993. Probing Soil and Aquifer Material Porosity with Nuclear Magnetic Resonance. Water . Resour.Res. 29:3861-3866.
(2) Provencher, S.W. CONTIN: A general purpose constrained regularization program for inverting noisy linear algebraic and integral equations. 1982. Comput.Phys.Commun. 27:229-242.

A Rainfall Soil Erosion Model

W.L Hogarth[1], C.W. Rose[1], G. Sander[2], I. Lisle[1], P. Hairsine[3], and J.Y. Parlange[4]. [1]*Faculty of Environmental Sciences, Griffith University, Brisbane, Australia.* [2]*Faculty of Science and Technology, Griffith University, Brisbane, Australia.* [3]*Division of Soils, CSIRO, Canberra, Australia.* [4]*Department of Agricultural Engineering, Cornell University, Ithaca, NY14583 USA.*

Introduction. In recent years there has been a number of alternative quantitative methods developed to describe the process of soil erosion by water on hillslopes. In most of these there is recognition that both rainfall impact and overland flow play a role when relatively shallow overland flow is dominant. Hairsine and Rose (1,2,3) developed a steady state theory to describe this situation. Their approach was to consider the contributions of the individual particle size classes to the total suspended sediment concentration. This lead to separate coupled conservation equations for the suspended sediment $c_i(x,t)$ for each particle size i. Prediction of size distributions of eroded particles enables estimates to be made of quantities of sorbed pollutants (4). An understanding of the dynamics of the suspended sediment concentrations c_i of the individual particles, as well as the overall total concentration c, therefore has significant implications on the understanding of the supply of non-point source pollutants to waterways. In this paper we outline a dynamic theory of rainfall impact-driven erosion. The theory enables description of the behaviour of the individual class sizes of soil particles both in the suspended sediment and the layer of deposited soil which develops during the erosion process. A comparison is made between the theory and experimental data from a tilting flume subject to simulated rainfall.

Methods. Let v_i denote the settling velocity of the sediment concentration c_i. Denoting the rate of deposition per unit area by d_i, then $d_i = v_i c_i$ i=1,2,...I where I is the number of size classes. The settling velocity classes v_i are chosen so that there is an equal mass of soil in each class. Since deposition returns sediment to the soil bed, then at any given time some fraction H of the original soil surface is shielded by the deposited layer of sediment. During erosion by rainfall the rate of detachment of soil particles e_i is given by

$$e_i = aP\,[I-H\,(x,t)]/\,I, \qquad\qquad i = 1,2,- I \qquad\qquad [1]$$

where a is the rainfall detachability of bare soil and P the rainfall rate. The redetachment rate of size class i, e_{di}, by rainfall of sediment in the deposited layer is given by

$$e_{di} = a_d\,P(M_{di}/M_{dt})\,H, \qquad\qquad i = 1,2,- I \qquad\qquad [2]$$

where M_{di} is the mass fraction of particles in that size class, a_d is the detachability of the deposited layer is given by

$$M_{dt} = \sum_{i-1}^{I} M_{di} \qquad\qquad [3]$$

The fractional coverage H(x,t) provided by the deposited layer, is expressed as

$$H\,(x,t) = M_{dt}/M_{dt}^* \qquad\qquad [4]$$

where M_{dt}^* is the mass per unit area of sediment required for complete average of the original soil. If D(x,t) is the depth of flow, then conservation of mass of sediment of size class i on a plane land element of uniform slope requires that

$$\partial(Dc_i)/\partial t + \partial(qc_i)/\partial x = e_i + e_{di} - d_i \quad i = 1,2,- I \qquad\qquad [5]$$

We assume kinematic overland flow with volumetric water flux given by $q = KD^m$ and mass conservation of water requires that

We assume kinematic overland flow with volumetric water flux given by $q = KD^m$ and mass conservation of water requires that

$$\partial D/\partial t + \partial q/\partial x = R(t) \qquad [6]$$

where $R(t)$ is excess rainfall. Conservation of mass in the deposited layer requires that

$$\partial M_{di}/\partial t = d_i - e_{di} \qquad i = 1,2, -, I \qquad [7]$$

Hence there are $2I$ partial differential equations to be solved for c_i and m_{di}.

Results and Discussion. The result of model simulation is compared with the experimental data of Profitt et al [5]. Figure 1(a) shows the model provides an excellent fit to the data at the end of the flume ie $x = L$. We also note the rapid rise of the sediment concentration followed by the decline towards its steady state value. In addition the behaviour of the sediment concentration with time can be considered as a function of distance down the flume. Figure 1(b) illustrates this for constant water depth and no inflow at the top of the flume. The model also provides the opportunity to explore characteristics of individual size classes both in the sediment concentration and the deposited layer.

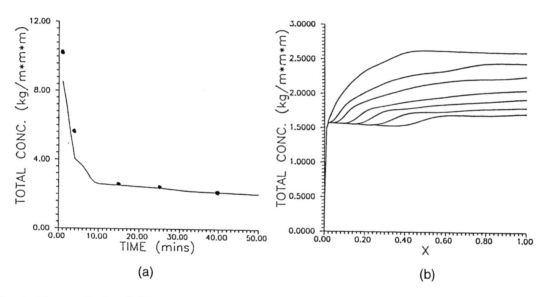

(a)   (b)

Fig. 1. The vertisol soil Blackearth is considered when P = 100 mm/hr and D = 5 mm (a) Sediment concentration as a function of time is given at $x = L$. Solid line is model, dots represent experimental data. (b) Sediment concentration as a function of x (nondimensional). Solid lines represent model prediction for the times ranging from 10 mins (upper curve) through 20,30,50,70,100,140 mins (lowest curve).

Literature Cited.
(1) Hairsine, P.B., and C.W. Rose. 1991. Rainfall detachment and deposition: Sediment transport in the absence of flow-driven processes. Soil Sci. Soc. Am. J., 55: 320-324.
(2) Hairsine, P.B. and C.W. Rose. 1992. Modelling water erosion due to overland flow using physical principles 1. Sheet flow. Water Resour. Res. 28: 237-243.
(3) Hairsine, P.B. and C.W. Rose. 1992. Modelling water erosion due to overland flow using physical principles 2. Rill flow. Water Resour. Res. 28: 245-250.
(4) Palis, R.G., G. Okwach, C.W. Rose and P.G. Saffigna. 1990. Soil erosion processes and nutrient loss. 1. The interpretation of enrichment ratio and nitrogen loss in runoff sediment, Aust. J. Soil Res. 28: 623-639.
(5) Profitt, A.P.B., C.W. Rose and P.B. Hairsine. 1991. Rainfall detachment and deposition: experiments with low slopes and significant water depths. Soil Sci. Soc. An. J. 55: 325-332.

Prediction of Wetting Front Movement During Infiltration into Soils

C. S. Kao and J. R. Hunt. *Department of Civil Engineering, University of California, Berkeley, CA 94720.*

Introduction. Efficient and accurate methods of predicting infiltration processes in both dry and moist soils are needed because of the prevalence of contaminants placed intentionally or accidentally near the soil surface. Most models that have been proposed are based on numerical or analytical solutions of Richard's equation and are mathematically complex or computationally expensive. The accuracy of these models is limited by uncertainties in the values of the hydraulic parameters needed to obtain solutions. Simpler models based on easily determined fluid and porous medium properties may have predictive capabilities that rival those of more complicated models. We present a wetting front model for fluid infiltration into porous media derived from the original work of Green and Ampt. The model is compared with published data and with data measured in controlled horizontal infiltration experiments.

Theory. Approximating the wetting front advance as a plug flow process allows soil to be modeled as a bundle of capillary tubes during infiltration. Using Poiseuille's law and the relationship between permeability and the square of a characteristic pore radius, we derive the following expression for horizontal infiltration into initially dry soil with a constant zero source pressure:

$$x_f = B \left(\frac{\sigma}{\mu} \right)^{\frac{1}{2}} k^{\frac{1}{4}} \sqrt{t} \qquad [1]$$

where x_f is the distance to the weting front, σ is the air-liquid interfacial tension, μ is the liquid viscosity, k is the intrinsic permeability of the porous medium, and t is the time. The coefficient B reflects the porous media geometry and should be constant for geometrically similar porous media. Data presented in the literature indicate that B has a value close to 0.5 for a wide range of soil types.

This relationship can be extended to cases of initially moist soil and negative source pressures and holds for infiltration of both water and nonaqueous phase wetting liquids. An expression for the pressure head at the wetting front can be obtained by equating equation (1) with the Green and Ampt horizontal infiltration equation, and the derived relationships can be combined with the Green and Ampt and Philip models to predict vertical infiltration. Data from vertical infiltration experiments described in the literature compare well with model predictions.

Material and Methods. We performed controlled horizontal infiltration experiments on glass beads, desert soil from the Nevada Test Site, and floated silica flour. The three soils spanned four orders of magnitude in permeability. The soil was packed into 1.67 cm diameter, 50 to 60 cm long, acrylic cylinders connected to a constant head reservoir. The infiltrating liquids used were silica oil at three different viscosities and water. The liquid source was held at atmospheric pressure, and the wetting front location and quantity of liquid imbibed were visually monitored with time. Experiments were performed on initially air dried and initially moist soil. Electrical resistivity probes were used to monitor water movement during infiltration of silica oil into initially water moist soil.

Results and Discussion. The agreement of the model with measured data is fair to good for infiltration under a zero source pressure. Figure 1 compares model predictions with data measured from our experiments for horizontal infiltration of silica oil into desert soil, and Figure 2 compares the model predictions with data taken from vertical infiltration experiments performed by Youngs and Price (1). Predictions for infiltration under a negative source pressure are sensitive to relatively small errors (a factor of 2) in estimates of permeability and wetting front suction head.

Conclusions. The proposed model shows potential for being a reliable and convenient method of estimating the extent of horizontal and vertical infiltration under non-negative inlet heads. Finer details, such as the variation in moisture content in the wetted region, cannot be predicted and must be obtained by actual measurements. However, measurements are necessary to check even the most precise numerical models because subsurface uncertainties preclude highly accurate modeling results. The strengths of the proposed model lie in its minimization of parameters that are difficult to measure, its synthesis of established ideas, and its emphasis on the most pertinent concepts to obtain predictions more easily and quickly than more complex models.

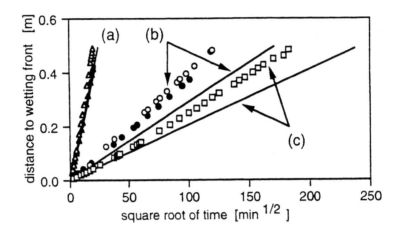

Figure 1. Data (symbols) and predictions (Lines) for horizontal infiltration of silica oil into initially air-dried desert soil under zero source pressure. The silic oil had viscosities of (A) 8.14×10^{-4} N-s/m^2, (b) 0.0479 N-s/m^2, and (c) 0.0961 N-s/m^2.

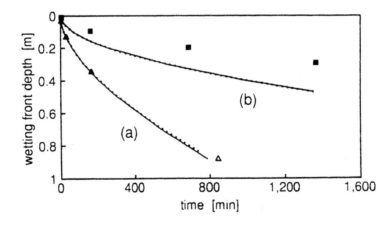

Figure 2. Data (symbols) measured by Youngs and Price (1) and predictions with the modified Green-Ampt (lines) and Philip (dotted lines) formulations for vertical infiltration of water into (a) slate dust, and (b) Rothamsted silt loam under zero source pressure.

Literature Cited.

(1)   Youngs, E. G., and R. I. Price. 1981. Scaling of infiltration behavior in dissimilar porous materials. Water Resour. Res. 17(4):1065-1070.

An Approach to Hysteresis Using Similarity

F.Kastanek. *Institute for Hydraulics and Rural Water Management, University of Agriculture, Forestry and Renewable Resources Vienna, Austria.*

Introduction. The hysteresis of the soil water characteristic is still a fascinating problem with strong practical impacts in modeling soil water behaviour. Hysteresis has important effects on water and solute distributions during field conditions that involve alternative wetting and drying (1). There are several efforts to describe hysteresis effects, but two of them are used more extensively: the method proposed by (2) using dependent domain theory and the method proposed by (3), revised by (4), using similarity. But there are more possibilities to describe similarity.

Methods. The following equations are proposed to estimate hysteresis:
For drying relationships (Fig. 1a):

$$\frac{HD-H}{HD-HS} = \left(\frac{W-WR}{WO-W} \frac{WO-WHS}{WHS-WR} \right)^{\alpha_1}$$ [1a]

or

$$\frac{WD-W}{WD-WS} = \left(\frac{HR-H}{H-HO} \frac{HHS-HO}{HR-HHS} \right)^{\alpha_2}$$ [1b]

For wetting relationships (Fig. 1b):

$$\frac{HD-H}{HD-HS} = \left(\frac{W-WHS}{W-WR} \frac{WO-WR}{WO-WHS} \right)^{\alpha_3}$$ [2a]

or

$$\frac{WD-W}{WD-WS} = \left(\frac{HHS-H}{HR-H} \frac{HR-HO}{HHS-HO} \right)^{\alpha_4}$$ [2b]

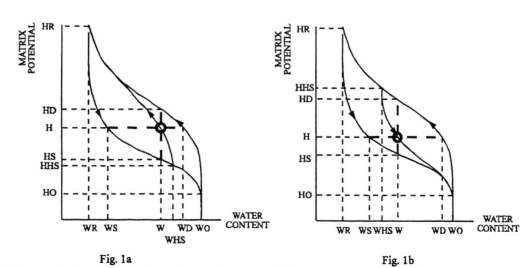

Fig. 1a   Fig. 1b

Fig. 1: Schematical representation of :   a.)drying scanning curve   b.)wetting scanning curve
They show good agreement with measured data, but the exponents used are not constant and no relationships with other parameters could be found. For simplicity the exponents were assumed to be unity

They show good agreement with measured data, but the exponents used are not constant and no relationships with other parameters could be found. For simplicity the exponents were assumed to be unity and by combining equations [1a] and [1b] and combining partially equations [2a] and [2b] some more equations could be found which fit given data much better and which can be used if the main drying and the main wetting relationships are known:

For drying relationships:

$$\frac{HD-H}{HD-HS}\frac{WD-W}{WD-WS} = \frac{W-WR}{WO-W}\frac{WO-WHS}{WHS-WR}\frac{HR-H}{H-HO}\frac{HHS-HO}{HR-HHS} \quad [3]$$

for wetting relationships:

$$\frac{HD-H}{HD-HS}\frac{WD-W}{WD-WS} = \frac{W-WHS}{W-WR}\frac{WO-WR}{WO-WHS} \quad [4]$$

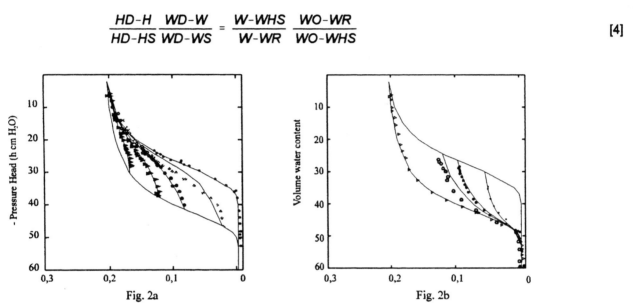

Fig. 2: Measured (Gilham et. al., 1976) and with the proposed method calculated scanning curves: a.) drying scanning curves;   b.) wetting scanning curves

Results and Discussion. In Fig. 2a and Fig. 2b measured data given by (5) are compared with data calculated with equations [3] and [4]. Comparisons with other measured data show similar good agreements. The application of the proposed new method using similarity needs the knowledge of the main drying and the main wetting relationships. To solve the equations [3] or [4] an iterative procedure is needed because WD and WS are dependent on H; and HD and HS are dependent on W, but the convergence is rather good.

Literature Cited.
(1)   Nielson D.R., M.Th.van Genuchten, J.W.Biggar 1986. Water flow and solute transport processes in the unsaturated zone. Water Resour. Res., 22:89S-108S.
(2)   Mualem Y. 1984. A modified dependent-domain theory of hysteresis. Soil Sci. 137:283-291.
(3)   Scott P.S., G.J.Farquhar, and N.Kouwen 1983. Hysteretic effects on net infiltration. Advances in Infiltration. American Society of Agricultural Engineers. ASAE Publication 11-83:163-170.
(4)   Kool J.B. and J.C.Parker 1987. Development and evaluation of closed-form expressions for hysteretic soil hydraulic properties. Water Resour.Res. 23:105-114.
(5)   Gilham R.W., Klute A. and D.F. Heermann 1976: Hydraulic properties of a porous medium: measurement and empirical representation. Soil Sci. Soc. Am. J. 40: 203-207.

Soil Characterization for Spatially Explicit Regional Hydrological Modeling

J.S. Kern[1], D. Marks[2], and K. Novins[3] . [1] *ManTech Environmental Technology Inc.,* [2] *US Geological Survey,* [3] *Oregon State University; all at the US Environmental Protection Agency National Health and Ecological Effects Laboratory - Western Ecological Division, 200 SW 35th Street, Corvallis, OR 97333 USA.*

Introduction. Modeling the soil water balance on a spatial basis over large areas for global change studies requires more simplification than modeling at a site. Modeling at regional scales provides challenges both for simplifying the processes involved and simplifying the spatial distribution of soil water properties. The purpose was to develop tools to study large river basins, such as the Columbia River, at a spatial resolution of 1-km and a daily time step. At such spatial and temporal scales detailed approaches for water flow such as the Richards' equation (1) are not practical because of the computing and disk storage resources required, as well as the lack of detailed input data. Furthermore, such approaches are not always necessary because information about soil moisture may only be needed on a daily or monthly basis. In this study a simple model was compared to an existing complex model to assess the effects of simplification. In addition, the effects of different approaches of describing soil hydraulic properties on a spatial basis were compared.

Materials and Methods. An existing complex model of water flow was compared to a simple approach for idealized sites of different soil textures. The complex model, the SMWS_2D, is based on the Richards' equation for saturated and unsaturated water flow (1). A simple model was developed by assuming that soil moisture reached -10 kPa soil water pressure after each day if the soil water pressure was greater than -10 kPa, and that it reached -33 kPa soil water pressure if the soil water pressure was greater than -33 kPa and less than -10 kPa. If the soil water pressure was below -33 kPa only evapotranspiration reduced soil moisture. Infiltration occurred until saturation water content was reached. The minimum soil water content was assumed to occur at -1500 kPa soil water pressure. Soil water properties were estimated from physical properties using pedotransfer functions (2,3). The effects of describing soil water properties spatially by area weighting averaging, and using multiple components was studied using spatial soil data from the USDA Natural Resource Conservation Service's state soil geographic database (STATSGO) (4) for the western USA. Area weighted averaging of soil texture, bulk density, organic matter, rock fragments and soil depth before using pedotransfer functions was compared with area weighted averaging of the soil water properties calculated for each map unit component separately. Map unit components were evaluated for their area of organic soils, drainage, texture, rock fragments, depth, and organic matter content to determine the number of hydrologically significant soil groupings.

Results and Discussion. The simplified soil water balance model approximated the detailed model after several days of model simulation although some fine and coarse textures deviated from -33 kPa soil water pressure. One reason for the model agreement was that the daily precipitation data that were available do not describe storm intensity. Thus, short, intense rainfall events that would result in different infiltration and flow among the models was not simulated. This brings out the point that even if it were more feasible to implement the detailed model, available input data are not always appropriately detailed. Area weighted averaging of soil properties before application to pedotransfer functions resulted in larger variability for water retention than first analyzing each map unit component separately. The variability of spatially averaging hydraulic conductivity was large due the nonlinear response of the pedotransfer functions (3) to texture. In the western USA, organic soils were not extensive and soil depth, rock fragment content, drainage, and soil depth often differentiated soil groups that were hydrologically significant. On-going work is being conducted to develop ways to lump hydrologically significant soil components into groups based on their water retention at soil water pressures of -10, -33, and -1500 kPa.

Conclusion. Simplified approaches to modeling the soil water balance for long time steps can be effective. Detailed models can provide information at time steps that are not needed for some applications. The detailed models may not be appropriate when input parameters are themselves generalizations of more detailed data. It is better to use pedotransfer functions for data for each map unit component rather than using the average soil properties of the components. The simple model has the advantage of not requiring hydraulic conductivity that is very difficult to estimate from basic soil properties (5) and is difficult to average spatially. The relatively large number of hydrologically significant soils observed here and the occurrence of components with properties that are that are difficult to average together suggests that it is better to characterize the map units with more than one soil grouping. The simplification of soil water balance models and the lumping of continuous pedotransfer functions by similar texture can conserve computing power and disk storage to the point where the implementation of multiple soil components can be feasible.

Literature Cited.

(1) Šimůnek, T. Vogel, and M.Th. van Genuchten. 1994. The SWMS_2D code for simulating water flow and solute transport in two-dimensional variably saturated media, ver. 1.21. Research Report No. 132, US Salinity Laboratory, USDA-Agricultural Research Service, Riverside, CA.

(2) Rawls, W.J., D.L. Brakensiek, and K.E. Saxton. 1982. Estimation of soil water properties. Trans. ASAE 25:1316-1320.

(3) Saxton, K.E., W.J. Rawls, J.S. Romberger, and R.I. Papendick. 1986. Estimating generalized soil-water characteristics from texture. Soil Sci. Soc. Am. J. 50:1031-1036.

(4) Soil Conservation Service. 1993. State soil geographic data base (STATSGO): data users guide. Miscellaneous Publication Number 1492. US Government Printing Office, Washington, DC.

(5) Kutlick, M. and D.R. Nielsen. 1994. Soil hydrology. Catena Verlag, Crelingen-Destedt, Germany.

Chloride and Water Content in the Root Zone of Barley Grown Under Four Salt-Water Irrigation Regimes

M.B. Kirkham[1], in collaboration with **Don Kirkham.**[2] [1]*Dept. of Agronomy, Kansas State Univ., Manhattan, KS 66506.* [2]*Dept. of Agronomy, Iowa State Univ., Ames, IA 50011.*

Introduction. Miller, Biggar, and Nielsen (1) showed that chloride movement in soil depends upon the method of water application. They found that intermittently ponding the soil with 51-mm (2-inch) increments of water was more efficient in leaching applied chloride from the soil surface than continuous ponding or leaching intermittently with 152-mm (6-inch) increments. For example, after 12 inches (305 mm) of water had been applied in 6- or 2-inch increments, the Cl⁻ concentration at the 152-mm depth was 75 and 20 mmol/L, respectively. They worked on plots with no vegetation. Our experiment was done to determine the effect of irrigation regime on chloride distribution and growth of barley. The hypothesis was that plants would grow better with small, frequent irrigations because Cl⁻ concentrations would be less in the soil surface than with large, infrequent irrigations.

Materials and Methods. Barley (*Hordeum vulgare* L.) grew in 48 large (280 mm tall; 280 mm diameter) clay pots, 24 containing fine white sand and 24 containing soil (2 parts loam:1 part humus:1 part coarse sand). Pots were thinned to 12 plants 25 d after planting, when differential salt and irrigation treatments began. The 3 salinization treatments were: 1) watering with a 0.39 *M* NaCl solution (23,000 ppm) (the concentration of sea water is about 33,000 ppm), 2) watering with a 0.18 *M* NaCl solution (10,500 ppm), and 3) watering with a 0 *M* NaCl solution. The 4 salt-water irrigation treatments were: 1) 250 mL/1 d, 2) 1000 mL/4 d, 3) 2000 mL/8 d, and 4) 3000 mL/12 d. Thus, every pot got the same amount of solution, but at different intervals. The NaCl added to the soil was dissolved in tap water, and the NaCl added to the sand was dissolved in nutrient solution. Chloride concentration of sand and soil samples taken from the pots at 51-mm depth increments to a depth of 254 mm was determined 8 times during the experiment: 4 times for Replicate I (days 16, 37, 48, and 58 after planting) and 4 times for Replicate II (days 31, 41, 53, and 68 after planting). Both chloride and moisture content (gravimetric analysis) were determined at the 0, 102, and 203 mm (0, 4, 8 inch) levels. Only chloride was determined at the 51, 152, and 254 mm (2, 6, 10 inch) levels. Pan evaporation in the greenhouse averaged 2 mm/d during the experiment (from planting on 10 Jan. to harvest on 18 April). In the figures, chloride and moisture data for each depth have been averaged over all times (8 sampling dates) to give a mean concentration in the pots throughout the experiment.

Results and Discussion. To conserve space, only Cl⁻ and moisture data for the 0.39 *M* NaCl under 3 irrigation regimes (250 mL/d, 1000 mL/4 d, 3000 mL/12 d) will be given. Irrigation frequency did not affect Cl⁻ distribution with depth in the sand (Fig. 1). In the soil, Cl⁻ concentration in the surface (0-51 mm) increased as the amount of solution added at each irrigation increased (Fig. 1), which agrees with Miller et al. (1). But dry weight increased as the irrigation frequency decreased (Table 1). In the soil, yield at harvest of plants irrigated with 0.39 *M* NaCl at a rate of 250 mL/1 d was 0 (a lethal combination), but growth occurred with 1000 mL/4 d. This was due, in part, to the higher water content in the surface of the soil when plants were irrigated less frequently (every 4, 8, or 12 d) (Fig. 2), a finding also observed by Miller et al. (1). Apparently, when the soil was allowed to dry between irrigations, the water (and dissolved Cl⁻) was drawn by capillarity into more of the soil's pore space in the surface where it was available for plant growth. When water was added daily, much must have moved to depth by gravity, because capillary attraction at the surface was less (2). The concentration of Cl⁻ was higher in the soil than sand, yet plants grew in the soil, because the moisture content was higher in the soil. The results showed that for growth under saline conditions, irrigations should be infrequent (e.g., 1000 mL/4 d).

Table 1. Shoot dry weight (g/pot ± SD) at harvest of barley irrigated with NaCl.

Irrigation frequency	Sand	Soil	Sand	Soil	Sand	Soil
	0.39*M*		0.18*M*		0.00*M*	
250 mL/ 1 d	0.0±0.0	0.0±0.0	5.0±0.0	11.0±4.2	39.0± 1.4	49.5± 9.2
1000 mL/ 4 d	0.0±0.0	0.3±0.4	3.3±1.1	13.0±0.0	30.5±13.4	58.0±14.1
2000 mL/ 8 d	0.0±0.0	0.8±0.4	1.8±0.4	10.0±1.4	38.0± 7.1	45.0±11.3
3000 mL/12 d	0.0±0.0	0.8±0.4	0.3±0.4	10.3±1.8	63.0±24.0	67.3± 3.9

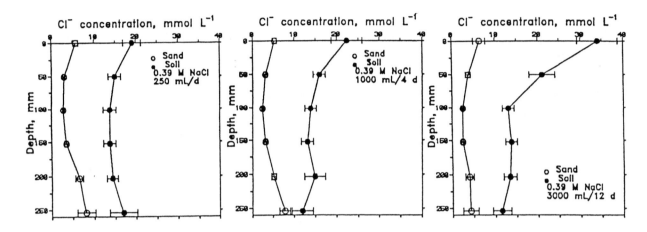

Fig. 1. Chloride distribution (± SE).

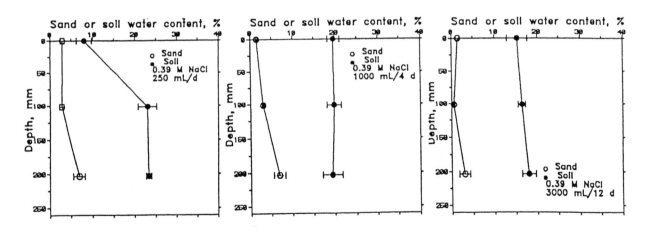

Fig. 2. Gravimetric moisture content (± SE).

Literature Cited.
(1) Miller, R.J., J.W. Biggar, and D.R. Nielsen. 1965. Chloride displacement in Panoche clay loam in relation to water movement and distribution. Water Resour. Res. 1:63-73.
(2) Clothier, B.E., and S.R. Green. 1994. Rootzone processes and the efficient use of irrigation water. Agr. Water Manage. 25:1-12.

Flow of Water in Layered Soils: Hypothesis on Release of Air

M. Kutilek. *Czech Technical University, 160 00 Prague 6, Czech Republic.*

Introduction. In some experiments on steady infiltration into crust topped soil columns, an additional resistance was observed in the vicinity of the boundary between the crust and the soil below (1,2). Zayani (3) performed unsteady infiltration experiments with ceramic plate simulating the action of the crust on sandy columns. He measured the increase of the hydraulic resistance of the plate after infiltration experiments. Two papers published in Japanese according to (4) discuss the release of air from soil water together with the observed decrease of hydraulic conductivity. I suppose that in crust topped soils and in layered soils the steep drop of the water pressure with depth may stimulate the release of air which could form microbubbles blocking some pores and consequently, hydraulic resistance would increase. The aim of the paper is to check experimentally on macroscopic scale the hypothesis on the role of Henry's law in flow of water in non-homogeneous soils.

Formulation of the Problem. In the first approximation, let us consider 1-d steady flux density q. For the concentration of dissolved air c and for the amount of released air a is

$$-q \, \frac{\partial c}{\partial x} = \frac{\partial (a+c)}{\partial t}$$ [1]

where x is the coordinate and t is time. Taking $\partial c/\partial t$ as negligible, we get with

$$\frac{\partial a}{\partial t} = \beta(c - \alpha a) \qquad a(x,0) = 0, \quad c(0,t) = c_o$$ [2]

after differentiating (1) with respect to t

$$q \, \frac{\partial^2 a}{\partial x \partial t} + \beta \, \frac{\partial a}{\partial t} + \alpha\beta \frac{\partial a}{\partial x} = 0$$ [3]

with

$$a(x,0) = 0 \quad and \quad a(0,t) = \frac{c_0}{\alpha}[1 - \exp(-\alpha\beta t)]$$ [4]

where α, β are coefficients of kinetics and air release in linear relationships. Then with a(x,t) we expect that K(x,t) where K is "saturated" hydraulic conductivity. Therefore, we get q(t) even for a constant pressure gradient and (1) should be corrected accordingly if solution of a(x,t) is searched.

Materials and Methods. Crust topped soil was modeled in the traditional way by ceramic plate of high resistance placed on the top of the homogeneous sand column. Thickness of the plate was L = 0.74 cm, thickness of the sand column was L = 10.10 cm. Hydraulic resistance R = L/K was for sand column R = 0.33 h, constant in time and for all delta H. Here, delta H is the pressure head difference at the ends of inflow and outflow of the tested material. Plate resistance R = 28.3 h for delta H = 3 cm was constant in time. However, R was higher for higher delta H, rising with time and reaching a quasi constant highest value after large time. Several sets of experiments were performed with variation of saturation of the plate and use of water either after, or without de-airation. R of the plate was measured during and after infiltration either separately or as the part of water saturated system plate-sand.

Results and Conclusions. Hydraulic resistance of the plate depends upon the procedure of saturation and the type of experimentation. It was rising with time during infiltration and during the saturated flow in the system plate-sand. Even in a short period of infiltration (20 min) R of the de-aired plate rose from R = 28.3 h to R = 82.3 h and for the non de-aired plate from R = 60 h to R = 77 h in one type of experiment, or from R = 83 h to R = 106 h in an other type of experiment. In the saturated flow in the plate-sand system, the time dependence R(t) of the plate at delta H = const. resembles the exponential relationship. The change of R is only partly reversible when the response of R to delta H is tested. The decrease of delta H from higher values to lower ones is accompanied by only partial decrease of R which never reaches its original low values. Low values of R of the plate were regained only after submersion of the plate in water at partial vacuum for long time. The measured data are in agreement with the assumed influence of soil water pressure drop upon the exclusion of the dissolved air. I obtained similar results with a system in sandwich arrangement plate-silt-plate. However, the content of air in the plate was not measurable by classical instrumentation and the blocking effect of microbubbles of air cannot be quantitatively interpreted by existing models.

Literature Cited.
(1)   Srinilta, S.A., D.R. Nielsen, and D. Kirkham. 1969. Steady flow of water through a two-layer soil. Water Resour. Res. 5:1053-1063.
(2)   Kutilek, M. 1974. Steady infiltration into soil pro file with a crust (in Czech). Vodohospodarsky cas. 22:535-548.
(3)   Zayani, K. 1987. L'infiltration dans les sols avec croute: Etudes experimentale, numerique et quasi-analytique. These de docteur ingenieur. Univ.Sci. et Med. de Grenoble. France.
(4)   Iwata, S., T. Tabuchi and B.P. Warkentin. 1995. Soil Water Interactions. 2nd Ed., M.Dekker, New York.

Capillary Pressure-Saturation Relationships for Two- and Three-Fluid Porous Media with Fractional Wettability

Feike J. Leij[1] **and Scott A. Bradford**[2]. [1]*U.S. Salinity Laboratory, USDA-ARS, Riverside, CA 92507.*

Introduction. There is considerable concern over the presence of immiscible organic compounds in the subsurface environment. The fate and movement of these compounds is simulated with increasingly sophisticated computer models for multiphase flow. These models require knowledge of the relationships between capillary pressure (P_c) and saturation (S) for all contiguous fluids; in our case a water (w), oil (o), and air (a) phase. The experimental determination of P_c-S curves can be cumbersome and they are estimated indirectly by scaling available P_c-S data (1,2). Scaling approaches and empirical models of P_c-S relations assume, among other things, that the medium is wetted by a single liquid (3,4). However, the solid phase of most natural porous media contains both oil and water wet sites. The wettability also greatly affects the permeability of porous media for different fluids. Despite the importance of wettability, soil physicists and hydrologists have for several reasons paid little attention to the effect of fractional wettability on the hydraulic properties of porous media. The purpose of this research is to elucidate the effect of fractional wettability by measuring P_c-S data of blasting sands with different oil and water wet fractions.

Theory. The capillary pressure across a fluid-fluid interface is generally defined as the difference between the nonwetting and wetting fluid pressures. This definition is not convenient for media with fractional wettability since there are no unique wetting and nonwetting fluids. We define the equilibrium capillary pressure as $P_{ow} = P_o - P_w$. According to Laplace's equation:

$$P_{ow} = P_o - P_w = \frac{2\sigma_{ow}}{R}\cos(\phi_{sow}) \qquad [1]$$

where σ_{ow} is the interfacial tension (N/m), R is an equivalent pore radius, and ϕ_{sow} is the contact angle. The wettability may be characterized by the contact angle (ϕ), which is defined by a horizontal force balance at the contact line between oil, water, and a flat solid (Young's equation):

$$\cos(\phi_{sow}) = \frac{\sigma_{so} - \sigma_{sw}}{\sigma_{ow}} \qquad [2]$$

Water wets the solid for $\phi_{sow} < 90°$, while oil wets the solid for $\phi_{sow} > 90°$. Different contact angles may occur when the wetting fluid is advancing (ϕ_{sow}^A) or receding (ϕ_{sow}^R). The above definition is impractical to determine contact angles for actual porous media. Values for ϕ_{sow} can be obtained by curve fitting P_c-S data according to Laplace's equation (5). The displacement of a nonwetting by a wetting fluid requires less work than the reverse process. The United States Bureau of Mines (USBM) Method quantifies wettability with the area under the P_c-S curve (A) according to:

$$I_{USBM} = \log(A_w / A_o) \qquad [3]$$

Note that $I_{USBM}=0$ for equal fractions of water- and oil-wet solids ($A_w = A_o$). In a three-fluid system, the wettability at the air-oil-water contact line is related to the coefficient of spreading for a drop of an intermediate (I) fluid on a wetting (W) fluid. This coefficient is defined as (6)

$$\Sigma_{I/W} = \sigma_{NW} - (\sigma_{NI} + \sigma_{IW}) \qquad [4]$$

A negative value for Σ_{IW} indicates that the intermediate fluid does not spread on the wetting fluid, while a positive value suggests that the intermediate fluid will spread on the wetting fluid.

Materials and Methods. Porous media with fractional wettability were obtained by combining different portions of blasting sands that were water wet (i.e., untreated) or oil wet after an organosilane (OTS) treatment. The porous medium contains 25% very coarse sand, 50% coarse/medium sand, and 25% fine sand. Soltrol 220 was used as oil. The equilibrium interfacial tensions, measured with a du Noüy ring tensiometer (7), were σ_{aw}=0.052 N/m (i.e., accounting for oil contamination), σ_{ao}=0.024 N/m, and σ_{ow}=0.026 N/m. Spreading coefficients were $\Sigma_{o/w}$=0.002 N/m and $\Sigma_{w/o}$=-0.054 N/m. Two- and three-fluid P_c-S relations were obtained according to the "Brooks method" using an automated procedure described by Bradford and Leij (5).

Results and Discussion. Two-fluid P_{ow}-S_w^{ow} curves, the superscript denotes the fluids present in the porous medium, were measured during primary water drainage in oil-water media with 0, 25, 50, 75, and 100% OTS sand. The area under the curves decreased as the fraction of OTS sand increases — less work is required to displace water by oil. Negative P_{ow} values were measured during water imbibition; water can only displace oil when a positive pressure is exerted on oil relative to water. The retention model of van Genuchten (3) was modified to describe the effective water saturation for both positive and negative capillary pressures:

$$P_{ow} + \gamma = \frac{1}{\alpha}\left[\left(\bar{S}_w^{ow}\right)^{n/(1-n)} - 1\right]^{1/n}$$ [5]

where n and α are empirical parameters, while the parameter γ was set equal to the magnitude of the lowest observed P_{ow}. Relevant parameters for describing P_w-S_w^{ow} data were obtained for the primary and main drainage, and the main imbibition curves of water. These include values for residual saturations (S_{rw}^{ow}, S_{ro}^{ow}) and γ, which follow directly from the observations, and α_{PD} (primary drainage), α_{MI} (main imbibition), α_{MD} (main drainage), and n (same for all three curves). The wettability was also quantified with the USBM wettability index, I_{USBM}. Measurements on three-fluid systems were carried out while incrementally adding oil to a medium containing air and a fixed amount of water. A comparison between the curves for different OTS fractions, shows that P_{ow} depends on S_w^{aow} but not on S_t^{aow}. Just as for two-fluid media, P_{ow} decreases with increasing OTS fraction. Furthermore, P_{ow} is always positive for the 25 and 50% OTS media. In contrast, P_{ow} is almost exclusively negative during water imbibition of the 75 and 100% OTS media implying that water acts mainly as intermediate fluid.

Summary and Conclusions. Two- and three-fluid P_c-S relations were obtained for media with an OTS fraction of 0, 25, 50, 75, and 100%. The experimental P_{ow}-S_w^{ow} curves were quite different for primary and main drainage of water. The main drainage curve has a lower P_{ow} for a given S_w^{ow}. The difference increased with hydrophobicity (OTS fraction). The model of van Genuchten was modified by adding a shifting parameter, γ, to the pressure values so that $P_{ow} + \gamma$ is always greater than or equal to zero. This model could describe P_{ow}-S_w^{ow} data with a regression coefficients for the goodness of fit in excess of 0.965. In three-fluid media with fractional wettability, the values of P_{ow} depend on S_w^{aow} but not on S_w^{aow}. Much of these and other results can be explained with the contrasting wetting/spreading behavior of water and oil. The untreated sand (water as wetting fluid) and oil (water as intermediate fluid) have different affinities to water. On the other hand, the OTS sand (oil as wetting fluid) and water (oil as intermediate fluid) have a similar affinity to oil.

Literature Cited
(1) Leverett, M. C. 1941. Capillary behavior in porous solids. Trans. AIME, 142:152-169.
(2) Demond, A. H., and P. V. Roberts. 1991. Effect of interfacial forces on two-phase capillary pressure-saturation relationships. Water Resour. Res. 27:423-437.
(3) van Genuchten, M. Th. 1980. A closed form equation for predicting the hydraulic conductivity of unsaturated soils. Soil Sci. Soc. Am. J. 44:892-898.
(4) Lenhard, R. J., and J. C. Parker. 1988. Experimental validation of the theory of extending two-phase saturation-pressure relations to three-fluid phase systems for monotonic drainage paths. Water Resour. Res. 24:373-380.
(5) Bradford, S., and F. J. Leij. 1995. Wettability effects on scaling two- and three-fluid capillary pressure-saturation relations. Env. Sci. and Technol. (in press).
(6) Adamson, A.W. 1990. Physical chemistry of surfaces. 5th ed. John Wiley, New York.
(7) du Noüy, P.L. 1919. A new apparatus for measuring surface tension. J. Gen. Physiol. 1:521-524.

Implications of Spatial Variability of Soil Physical Properties for Simulation of Evaporation at the Field Scale

Elisabet Lewan and Per-Erik Jansson. *Swedish University of Agricultural Sciences, Department of Soil Sciences, Box 7014, S-750 07 Uppsala, Sweden*

Introduction. Soil physical properties affect evaporation by two principle mechanisms: by governing the transport of water towards the soil surface or root surface and by storing water. Water transport properties may be especially important for evaporation from bare and sparsely vegetated soils, whereas water storage properties may be relatively more important on densely vegetated soil. Spatial variability in the different properties of such a system is the rule rather than the exception. A number of studies have quantified and described the spatial variability of different soil properties. The importance of taking into account spatial variability in soil and plant properties when simulating water dynamics has been demonstrated by several researchers. However, only a few studies have focused on the importance of this variability for simulation of either the areal mean or the spatial variation in evaporation and transpiration [1,2,3,4,5]. The aims of this study are: (a) to present variation in soil physical properties for a sandy soil, based on field and laboratory measurements; (b) to investigate the importance of this variation for simulation of evaporation for individual locations and for simulation of the average evaporation for these locations, based on a period of 19 days of continuos drying.

Materials and Methods. Soil water retention curves in 8 soil profiles on a sandy soil were determined by field measurements of soil water content and soil water tension made during 19 days of continuous drying using TDR and tensiometers. Evaporation from the different profiles was simulated using the retention curves and estimated unsaturated conductivity functions, meteorological data as driving variables and measured soil water matric potentials at 40 cm depths as the lower boundary condition. The simulations were made using a one dimensional physically-based model for water and heat flows in the soil, SOIL [6,7]. The Penman-Montieth equation was used to calculate evaporation rates from the soil surface. The soil surface resistance was estimated by an empirical function based on the soil water potential of the uppermost soil layer.

Results and Discussion. Good agreement was obtained between simulated and measured evaporation from different locations in the field. This demonstrates that the variation between locations in measured evaporation could be partly explained by the variation in independently determined soil physical properties. Areal mean evaporation obtained as the arithmetic mean of individual simulations was larger than the evaporation obtained using mean physical properties in one simulation. However, this difference was relatively small compared to the variation in simulated evaporation between individual locations.

Literature Cited.
(1) Peck, A.J., R.J. Luxmoore, and J.L. Stolzy. 1977. Effects of spatial variability of soil hydraulic properties in water budget modelling, Water Resources Research, vol. 13(2), 348-354.
(2) Sharma, M.L., and R.J. Luxmoore. 1979. Soil spatial variability and its consequences on simulated water balance, Water Resour. Res., 15(6), 1567-1573.
(3) Hopmans, J.W., and J.N.M. Stricker. 1989. Stochastic analysis of soil water regime in a watershed, J. Hydrol., 105, 57-84.
(4) Lascano, R.J., and J.L. Hatfield. 1992. Spatial variability of evaporation along two transects of bare soil, Soil Sci. Soc. Am. J., Vol. 56, 341-346.
(5) Braud, I., A.C. Dantas-Antonino, and M. Vauclin. 1995. A stochastic approach to studying the influence of the spatial variability of soil hydraulic properties on surface fluxes, temperature and humidity, Journal of Hydrology, 165, 283-310.

(6) Jansson, P-E., and S. Halldin. 1979. Model for the annual water and heat flow in a layered soil. In: S. Halldin (Editor), comparison of Forest and Energy Exchange Models. Int. Soc. For. Ecol. Modelling, Copenhagen, pp. 145-163.

(7) Jansson, P-E. 1994. The Soil water and heat model, Technical description, Swedish University of Agricultural Sciences, Dept. of Soil Sciences, Division of Hydrotechnics, Rep. 165, Uppsala, 72 pp.

Identification of Optimal Locations for Sampling Ground Water for Pesticides in the Mississippi Delta Region of Eastern Arkansas

H.S. Lin, H.D. Scott, and J.M. McKimmey. *Dept. of Agronomy, Univ. of Arkansas, Fayetteville, AR 72701, USA.*

Introduction.   Concerns over the finding of pesticides in wells in the Mississippi Delta region of eastern Arkansas generate the need to assess ground water vulnerability to pesticide contamination in that highly-productive agricultural region. With a large area of crops grown in the Delta and only limited financial resources available for chemical analyses by state and federal agencies, the question of where these agencies should begin to sample and monitor the ground water for pesticides is pertinent. To assist the selection of optimal locations in the Arkansas Delta for sampling ground water for pesticides (which is required in the Arkansas Agricultural Chemical Ground-Water Management Plan), this study was conducted to rank the areas in the region according to their relative contamination potential in the alluvial aquifer. Based on currently available digital data, a vulnerability map showing the spatial distribution of relative ground water vulnerability index was sought using a modified Agricultural DRASTIC model (1) in a GIS environment. Process-based simulation models were difficult to use in this large scale study owing to inadequate methods of measuring and/or estimating the necessary model input parameters (2). The vulnerable areas identified through this study can be targeted by planners and governmental agencies for ideas where to direct resources for further site-specific and pesticide-specific investigations.

Materials and Methods.   The Arkansas Delta region encompasses 28 counties and about 10 million acres. Approximately 68% of the area are in cropland with extensive use of pesticides. The shallow alluvial aquifer lies from near 0 feet to about 140 feet below the land surface. The aquifer is composed of coarse sand and gravel that grade upward to fine sand. Sedimentary layers of clay, silt, and fine sand overlie and confine the aquifer. Ground water vulnerability assessment conducted in this study consisted of two separate evaluations: a) sensitivity assessment, i.e., the physical landscape parameters indicating the ease of pesticide leaching from land surface to the aquifer, and b) probability assessment, i.e., the likelihood of having a certain amount and type of pesticide applied to a land surface. Those areas with congruent high aquifer sensitivity and high pesticide use-intensity were considered to have the most vulnerable ground water to pesticide contamination. Following the recommendation of the Arkansas Soil and Water Conservation Commission (3), the Agricultural DRASTIC model was used to evaluate the relative sensitivity of the alluvial aquifer to pesticide contamination. To better utilize the existing data, we made several modifications to the original model. Satellite imagery (1992 Landsat 5 thematic mapper) was employed to identify crop distribution in the region. Pesticide use-intensity was then estimated from land use/land cover and pesticide use survey information (4) to assess the probability of pesticide contamination. Relative ground water vulnerability index was expressed as a product of aquifer sensitivity index and pesticide-use probability index.

The GRASS GIS was used as a spatial data manager in this study. Calculations and/or estimations of the major parameters needed in the DRASTIC model included: a) the potentiometric contour lines were interpolated into a full surface using a regularized spline with tension algorithm (5). Depth to ground water was then determined by subtracting potentiometric surface values from elevation data; b) net recharge rate was calculated by USGS using the MUDFLOW model at a one-square-mile cell scale (6); c) a state-level soil map was used to depict the dominant soils in the region (county-level digital soil maps are not currently available). Soil matrix permeability classes from the SOILS5 database (National Soil Survey Center, Lincoln, NE) were adjusted to fabric permeability values (matrix plus macroscopic features) using soil structure information from soil profile descriptions. Effective permeability of a soil profile was then calculated using horizon-thickness-weighted value (7); d) elevation data was utilized by GRASS to generate a map of percent slope for the region; and e) thickness of the confining unit overlying the alluvial aquifer, instead of the original qualitative description of geological material, was used as the impact of the vadose zone in calculating the DRASTIC index.

Results and Discussion.   The areas most sensitive to pesticide contamination in the Arkansas Delta were found to distribute mostly along major streams in the region, such as the Arkansas, Mississippi, White, Cache, St. Francis, and Bayou Bartholomew rivers, except in the Grand Prairie areas and the west side of Crowleys Ridge. In these river basins, a combination of shallow depth to ground water, thin confining unit, permeable soils, and high recharge rate usually prevails, leading to high sensitivity for the alluvial aquifer to be

Cache, St. Francis, and Bayou Bartholomew rivers, except in the Grand Prairie areas and the west side of Crowleys Ridge. In these river basins, a combination of shallow depth to ground water, thin confining unit, permeable soils, and high recharge rate usually prevails, leading to high sensitivity for the alluvial aquifer to be contaminated by pesticides. It was also in many of these sensitive areas where large acreages of crops were grown, and the largest amounts of pesticides were used. Consequently, many areas along major streams in the region were most vulnerable. Areal extent of the crops grown in the vulnerable areas showed that wheat-soybean double-cropped fields generally had the highest likelihood to be most vulnerable. Among the seven factors considered in the DRASTIC model, the impact of the vadose zone and the depth to ground water appeared to be the dominant factors that controlled aquifer sensitivity in the region.

Non-uniform distribution of the vulnerability index in the region suggests that random sampling of wells would not be an efficient approach for monitoring pesticide contamination problem. There was about 11% area of the region having a relative high sensitivity index and about 13% area showing a relative high vulnerability index.

Uncertainty is pervasive in both spatial databases and computational schemes in ground water vulnerability assessment. As a result, all currently available ground water vulnerability assessments are inherently uncertain (8). Sources of uncertainty and/or possible error involved in this study included: a) temporal dynamics of the model parameters; b) spatial variability of soil and hydrogeological properties; c) limitations in some data calculation algorithms; d) uncertainty in the Agricultural DRASTIC model concept; e) possible error/fuzziness in pesticide use information; and f) uncertainty in crop identification from satellite imagery. These uncertainties thus require continued and improved efforts in ground water vulnerability assessment for the Arkansas Delta region. Because of the dynamic nature of the factors involved in assessing ground water vulnerability, we believe that such assessment is not a one-time task, but, rather, should be performed periodically. More detailed information should be obtained and/or site investigations should be made on the most vulnerable or sensitive areas before specific action can be taken.

Acknowledgments. Funding for this research from the Arkansas Soil and Water Conservation Commission is gratefully acknowledged. Thanks are due to Mr. David Poynter at USGS in Little Rock for providing various digital data needed in this study, to Dr. E.M. Rutledge for help in obtaining and interpreting soil survey data, and to Dr. J. V. Brahana for helpful discussion. We also thank Bruce Gorham and Kim Hofer for helping interpret satellite imagery used in this study.

Literature Cited.
(1) Aller, L., T. Bennett, J. Lehr, R.J. Petty, and G. Hackett. 1987. DRASTIC: A standardized system for evaluating ground water pollution potential using hydrogeologic settings. EPA-600/2-87-035. U.S. Environmental Protection Agency, Ada, OK.
(2) Rao, P.S.C., and R.E. Jessup. 1982. Development and verification of simulation models for describing pesticide dynamics in soils. Ecol. Model., 16:67-75.
(3) Arkansas Soil and Water Conservation Commission. 1991. Identification of ground-water vulnerability areas in Arkansas. Technical Report. Little Rock, AR.
(4) Arkansas Cooperative Extension Service. 1992. Insecticide recommendations for Arkansas (MP144), Recommended chemicals for weed and brush control (MP44), and Plant disease control handbook (MP154). Little Rock, AR.
(5) Mitasova, H. 1992. Surfaces and modeling. Grassclippings. 6:13-18.
(6) Mahon, G.L., and D.T. Poynter. 1993. Development, calibration, and testing of ground-water flow models for the Mississippi River Valley alluvial aquifer in eastern Arkansas using one-square-mile cells. USGS Water-Resources Investigation Report 92-4106. Little Rock, AR.
(7) Jury, W.A., W.R. Gardner, and W.H. Gardner. 1991. Soil physics. Fifth ed. John Wiley & Sons, Inc., New York.
(8) National Research Council. 1993. Ground water vulnerability assessment: predicting relative contamination potential under conditions of uncertainty. National Academy Press, Washington, D.C.

An Extended Transfer Function Model for Field-Scale Solute Transport

H.H. Liu and J.H. Dane. *Department of Agronmy and Soils, Auburn University, Alabama 36849-5412, U.S.A.*

Introduction. Modeling field-scale solute transport through the unsaturated zone has been an important research problem for many years. Because it is very difficult to obtain adequate data describing field-scale transport and hydraulic properties as input into two- or three-dimensional models, many researchers have used one-dimensional models to predict area-averaged solute transport in the unsaturated zone (9). The commonly used one-dimensional models are based on the deterministic convection-dispersion equation (CDE) (1, 10) or on the principles of stochastic-convective transport (4, 5, 7).

Theory. When steady-state water flow exists in a field, models based on the deterministic CDE (referred to ASCDE), can be characterized by (6, 8, 9):

$$\frac{E_z(t)}{E_L(t)} = \frac{z}{L} \qquad \frac{Var_z(t)}{Var_z(t)} = (\frac{z}{L}) \tag{1}$$

where $E_i(t)$ and $var_i(t)$ are, respectively, the mean and variance of the travel time of the tracer particles moving from the soil surface to depth i ($i = z, L$). The inherent assumption in this approach is that horizontal mixing has been well developed or that the water velocity and concentration are uniform across the field at a given depth.

In contrast, models based on the principles of stochastic-convective transport (referred to as TFM) have the following properties (6,9):

$$\frac{E_z(t)}{E_L(t)} = \frac{z}{L} \qquad \frac{Var_z(t)}{Var_z(t)} = (\frac{z}{L})^2 \tag{2}$$

In this approach, no mixing in the horizontal direction is assumed to occur.

During field-scale solute transport through the unsaturated zone, however, some degree of horizontal (lateral) mixing is likely to occur. For that reason we propose a simple "black box" model (referred to as ETFM) to describe field-scale solute transport, based on the assumption that solute transport can be characterized by:

$$\frac{E_z(t)}{E_L(t)} = \frac{z}{L} \qquad \frac{Var_z(t)}{Var_L(t)} = (\frac{z}{L})^{2a} \tag{3}$$

where a is an adjustable parameter. Analogously to (5), we developed a relation between the tracer particle time probability density function, f_i, and depth i, $i = L, z$:

$$f_z(t) = (\frac{L}{z})^a f_L[\frac{t - E_L(t)(\frac{z}{L} - (\frac{z}{L})^a)}{(z/L)^a}] \tag{4}$$

The area-averaged flux solute concentration at depth z can then be obtained from the probability density function at depth L:

$$c_z(t) = \int_0^t c_{in}(\tau)(\frac{L}{z})^a f_L[\frac{t - \tau - E_L(t)(\frac{z}{L} - (\frac{z}{L})^a)}{(z/L)^a}] d\tau \tag{5}$$

Materials and methods. The field experimental data of (3) were used to test the ASCDE, the TFM, and the ETFM. To predict solute transport with any of the three models, the travel time probability density function at a reference depth $z = L$ needs to be determined. Once $f_L(t)$ is determined, the solute concentration at a depth $z > L$ can be determined by

$$\frac{c_z(t)}{c_0} = M \cdot f_z(t)$$ [6]

where M is a constant and the travel time probability density function $f_z(t)$ is related to $f(t)$ through Eq.[4]. For the ASCDE and the TFM, $a = 0.5$ and 1.0 were used in Eq.[6] and [4], respectively, to predict solute transport for $z > L$. For the ETFM, a needed to be determined from concentration data at a depth $z > L$.

Results and Discussion. Comparison between experimental data and modeling results indicated that the ETFM, with an a value determined from relevant experimental data, generally provided better prediction for field solute transport than either the CDE or the TFM (see Fig. 1 as an example). The ETFM, however, requires additional experimental data to determine a. Another possible approach is to treat a as a universal constant, i.e., the same a value is used for predicting solute transport in any field, such that the ETFM and the TFM need the same experimental data for model calibration. Based on the experimental data of (2, 3, 9), the value of such a universal a factor was determined to be 0.76.

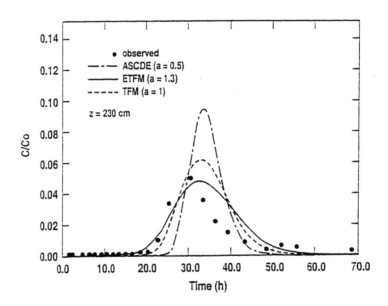

Fig. 1. Observed and predicted relative concentrations by three models.

Literature Cited.
(1) Biggar, J. W., and D. R. Nielson. 1967. Miscible displacement and leaching phenomena. Agronomy 11:254-274.
(2) Butters, E., W. A. Jury, and F. F. Ernst. 1989. Field scale transport of bromide in an unsaturated soil: 2. dispersion modelling. Water Resour. Res. 25:1583-1589.
(3) Costa, J.S., R.E. Knighton, and L. Prunty. 1994. Model comparison of unsaturated steady-state solute transport in a field plot. Soil Sci. Soc. Am. J. 58:1277-1287.
(4) Dagan, G., and E. Bresler, 1979. Solute dispersion in unsaturated heterogeneous soil at field scale. I. Theory. Soil Sci. Soc. Am. J. 43:461-467.
(5) Jury, W. A. 1982. Simulation of solute transport using a transfer function model. Water Resour. Res. 18:363-368.
(6) Jury, W. A., and K. Roth. 1990. Transfer functions and solute transport through soil: Theory and applications. Birkhaeuser Publ., Basel, Switzerland.
(7) Simmons, C. S. 1982. A stochastic-convective transport representation of dispersion in one

dimensional porous media systems. Water Resour. Res. 18:1193-1214.

(8)   Sposito, G., R. E. White, P. R. Darrah, and W. A. Jury. 1986. A transfer function model of solute transport through soil. 3. The convection dispersion equation. Water Resour. Res. 22:78- 88.

(9)   Tseng, P. H., and W. A. Jury. 1994. Comparison of transfer function and deterministic modelling of area-averaged solute transport in a heterogeneous field. Water Resour. Res. 30:2051-2063.

(10) van Genuchten, M. Th., and P. J. Wierenga. 1976. Mass transfer studies in sorbing porous media. I. Analytical solutions. Soil Sci. Soc. Am. J., 40:473-480.

Prediction of Field-Scale Transport Using a Stochastic Stream Tube Model

D. Mallants[1], N. Toride[2], M. Th. van Genuchten[3], and J. Feyen[1]. [1]*Institute for Land and Water Management, KULeuven, B-3000 Belgium.* [2]*Department of Agricultural Sciences, Saga University, Saga 840, Japan.* [3]*U.S. Salinity Laboratory, USDA, ARS, Riverside, CA 92501, USA*

Introduction. Field-scale solute transport is significantly influenced by the heterogeneous nature of flow and transport processes across the field. In the last two decades, both theoretical and experimental studies have been carried out to investigate solute behavior in field soils. These studies include investigations on the combined effects of pore-scale dispersion and velocity variability on solute distribution variations in the field (1), and the effects of local-scale transport parameters on field-scale transport for a variety of boundary and initial conditions (2). Experimental data obtained under well-controlled conditions are of utmost importance for assessing the predictive capabilities of theoretical models. Laboratory solute transport experiments using undisturbed soil columns provide useful information on the local-scale solute behavior. If field heterogeneity is accounted for by taking replicated samples, and if the particular flow conditions at the local scale are present also at the field scale, then the field-scale solute transport process can be readily evaluated by appropriate averaging of the local-scale processes.

The objectives of this paper were (1) to describe observed solute transport at the local scale using the classical one-dimensional convection-disperion equation (CDE), and (2) to assess the capacities of a stochastic stream tube model in predicting field-scale transport, thereby accounting for observed variability in local-scale pore-water velocity, v, and dispersion coefficient, D.

Materials and Methods. Solute transport experiments were carried out on 14 one-m long and 0.3-m diameter soil columns collected along a 31-m long transect in the field. Three different horizons were identified within the first meter of the soil profile, i.e. a Ap horizon (12.7% clay and 39.7% sand), a C1 horizon (16.6% clay and 24.1% sand), and a C2 horizon (21.8% clay and 20.1% sand). The soil, classified as an Udifluvent (Eutric Regosol), contained a large number of macropores throughout the profile (3). Twenty-five-cm long two-rod parallel TDR probes were inserted in the soil columns at six different depths. After establishing steady-state saturated flow, a solution of $CaCl_2$ was added to the soil and solute transport was observed by relating the TDR-measured bulk electrical conductivity of the soil, EC_a, to the average salt concentration in the soil, C (3). The step-type breakthrough curve (BTC) thus obtained served to estimate the dispersion coefficient from each BTC using a nonlinear inversion method based on the analytical solution of the CDE (4). Field-scale BTCs were obtained by averaging local-scale concentrations over the different columns. Field-scale transport was predicted using the stochastic CDE, which views the field as a set of homogeneous vertical soil columns (stream tubes) in which steady-state downward water flow occurs with no exchange between columns. The field-scale mean concentration, $<c>$, in terms of the stream tube model with stochastic v and D is given by (2):

$$<c(x,t)> = \int_0^\infty \int_0^\infty c(x,t;v,D)\, f(v,D)\, dv\, dD \qquad [1]$$

where $f(v,D)$ is the joint probability density function for v and D which may be described with the bivariate lognormal distribution:

$$f(v,D) = \frac{1}{2\pi\sigma_v\sigma_D vD\sqrt{1-\rho_{vD}^2}} \exp\left[-\frac{Y_v^2 - 2\rho_{vD}Y_v Y_D + Y_D^2}{2(1-\rho_{vD}^2)}\right] \qquad [2]$$

where

$$Y_v = \frac{\ln(v) - \mu_v}{\sigma_v}, \quad Y_D = \frac{\ln(D) - \mu_D}{\sigma_D} \qquad [3]$$

in which μ is the mean and σ the standard deviation of $\ln(v)$ and $\ln(D)$, and ρ_{vD} the correlation between Y_v and Y_D.

Results and Discussion. The relationship between the CDE parameters D and v was evaluated by using linear regression analysis to estimate the coefficients λ and n in the following equation:

$$D = \lambda v^n \qquad\qquad\qquad [4]$$

where λ is dispersivity (cm) and n a parameter generally ranging between 1 to 2. Combining all data leads to a 'field-scale' dispersivity of 31 cm and a value of 1.93 for the exponent n (Fig. 1a). This relatively high value for n reflects the relatively important effects of physical nonequilibrium on solute spreading. Of note in Fig. 1a is the relatively high variability in the $\ln(v)$ data for the Ap horizon as compared to the C1 and C2 horizons. This result indicates that flow had become more homogenous deeper in the soil profile, likely as a result of reduced biological activity.

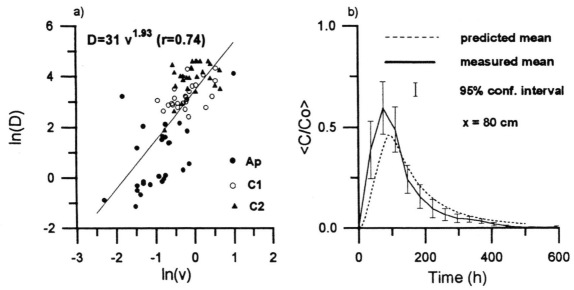

Fig. 1. (a) Relationship between v and D, (b) predicted and observed resident concentrations.

Field-scale transport at a depth of 80 cm was predicted upon application of eq. (1-3) and using the following transport parameters fitted to the 45, 60 and 80-cm depth data: $<v> = 1.02$ cm/h, $<D> = 56.4$ cm^2/h, $\sigma_{lnv} = 0.37$, $\sigma_{lnD} = 0.61$, $\rho_{vD} = 0.74$. The results (Fig. 1b) show a fairly good agreement between the mean measured and mean predicted BTCs. The location of the concentration peak is accurately predicted but the peak concentration itself is underestimated, whereas tailing is slightly overpredicted.

Literature cited
(1) Bresler, E. and G. Dagan. 1981. Convective and pore scale dispersive solute transport in unsaturated heterogeneous fields. Water Resour. Res. 17:1683-1693.
(2) Toride, N.., and F.J. Leij. 1994. Stochastic stream tube model for field-scale solute transport: 1. Theory. Soil Sci. Soc. Am. J. (submitted)
(3) Mallants, D., M. Vanclooster, M. Meddahi, and J. Feyen. 1994. Estimating solute transport in undisturbed soil columns using time-domain reflectometry. J. Contam. Hydrol. 17:91-109.
(4) Parker, J.C., and M. Th. van Genuchten. 1984. Determining transport parameters from laboratory and field tracer experiments. Bulletin 84-3. Virginia Agric. Exp. Sta., Blacksburg.

An Investigation of the Solution of Three-Dimensional Richards Equation using DASPK

D. Misra[1], J.L. Nieber[2], and R.S. Maier[1]. [1]*Army High Performance Computing Research Center, University of Minnesota.* [2]*Department of Biosystems and Agricultural Engineering, University of Minnesota.*

Introduction. The Richards equation is commonly used by scientists and engineers to predict the flow of water in an unsaturated porous media. The governing differential equation is of the following form:

$$\frac{\partial \theta}{\partial t} = \nabla \bullet K(h) \nabla h - \frac{\partial K}{\partial z} \tag{1}$$

where θ and h are volumetric moisture content and matric potential head, respectively; K is the unsaturated hydraulic conductivity; z and t are the space and time functions, respectively. Equation [1] has two dependent variables, θ and h. Hence, the soil water retention function is introduced in the form of a constitutive relationship in order to eliminate one of the variables from the solution process. As a result of this elimination, [1] is reduced to an equation with either 'h' or 'θ' as the dependent variable. Generally, the K-h and the θ-h relationships are highly non-linear. Hence, analytical solutions of [1] are scarce. However, numerical solutions of the Richards equation have been extensively investigated and reported in the literature. These numerical solutions have suffered from poor mass balances, unacceptable time step limitations, or poor CPU efficiency in the past (1). Several attempts have been made to improve the mass balance of the solution, however, the CPU efficiency and the time step limitations still remain a problem for certain conditions such as very dry initial conditions, extreme heterogeneities or hysteritic soils (1). This necessitates a highly CPU efficient, time-step adaptive computer code for modeling of unsaturated three-dimensional flow. Numerical methods for solving PDE's (such as the Richards equation) usually involve the replacement of all derivatives by discrete difference approximations. A typical example is the method of lines (MOL), where the approach is to discretize the spatial derivatives and thus convert the system of PDE's into an ODE initial value problem. The two advantages of the MOL approach are (a) it is computationally efficient and (b) the ODE software takes on the burden of time discretization and of choosing the time steps in a way that maintains accuracy and stability in the evolving solution. Most production ODE software is written to be robust and computationally efficient. Also, the person using a MOL approach has only to be concerned with discretizing spatial derivatives, thus reducing the work required to write a computer program. Many MOL problems lead to an explicit ODE. However, many well posed problems of practical interest are easily handled as a DAE (differential -algebraic equations) (6). The Richards equation in combination with the soil water retention function can be posed as a DAE when discretized spatially. The MOL can then be effectively used to obtain a robust and computationally efficient solution.

Methodology. DASPK (Differential Algebraic Systems, Preconditioned Krylov) solves large-scale systems of DAE. It is based on the integration method in DASSL (2), but instead of a direct method for the associated linear systems which arise at each time step, the preconditioned GMRES iteration (3) is applied in combination with an Inexact Newton Method. Two parallel versions of DASPK have been developed (4). DASPKF90 is a Fortran 90 data parallel implementation and DASPKMP is a message-passing implementation written in Fortran 77 with extended BLAS. The parallel versions have been implemented for the TMC CM-5, a massively parallel multiprocessor, keeping the user interface relatively simple while allowing for portability to other massively parallel architectures. The underlying idea for solving DAE systems (5) consists of replacing the solution and derivative of the DAE by difference approximation, and solving the resulting equation for the solution at the current time using Newton's method. Integration over the specified range of time is usually accomplished in a series of steps. The algorithm for computing an integration step involves replacing the derivatives with difference approximations and using a predictor-corrector method. In the predictor-corrector method an initial guess for the new solution is developed by evaluating the predictor polynomial, which interpolates solution values at previous time steps. The new solution is then computed more precisely by solving a nonlinear system of equations in the corrector step. The predictor and corrector polynomials are specified using the backward differentiation formulas of orders one through five. On every step, DASPK chooses the order and the stepsize based on the behaviour of the solution. In each corrector step a nonlinear system is solved by a Newton-type iteration. Each of these iterations requires the solution of a linear system. These linear systems are solved by the preconditioned GMRES. One of the powerful features of the iterative approach is that it does not need to compute and store the iteration matrix explicitly because the GMRES method never needs the matrix explicitly. For a mathematical description, readers are referred to (6).

Results and Discussion. The model has been validated using a test case from (7) comprising of a one-dimensional flow problem with a 200 cm deep soil column. The bottom boundary condition was a Dirichlet boundary condition with a specified pressure of -100 cm. The top boundary condition was a Neumann boundary condition with a specified flux of 0.9 cm/h. The initial condition was hydrostatic throughout the column. The domain comprised of two heterogeneous layers of equal depth with the top layer having a saturated hydraulic conductivity (K_s) of 10.0 cm/h and the bottom layer of 1.0 cm/h. The soil water retention and the unsaturated hydraulic conductivities are described using the equations:

$$K = K_s e^{\alpha\psi} \qquad\qquad [2]$$

$$\Theta = \Theta_r + (\Theta_s - \Theta_r) e^{\alpha\psi} \qquad\qquad [3]$$

The numerical values for the parameters used in the simulation are:
θ_s = 0.44 (saturated water content), θ_r = 0.067 (residual water content), and α = 0.01 cm^{-1}. The results obtained were in close agreement with the analytical solution for the problem. A three-dimensional, parallel, finite element code has been developed for the CM-5, based on the upwind scheme in (8) and using DASPKF90 for time integration (the validation case described above was actually solved as a three-dimensional problem). This computational testbed is currently being used to investigate the benefits of variable-order, variable-step time integration. Of particular interest is whether adaptive order of integration will permit significantly longer time steps during certain stages of a transient flow problem. Also of interest is the extent to which Richards equation becomes stiffer as the mesh is refined, and the resulting effect on integration time. Related questions include the parallel efficiency as a function of mesh refinement.

Literature Cited.
(1) Pan, L., and Wierenga, P.J., 1995. A transformed pressure head-based approach to solve Richards' equation for variably saturated soils, Water Resour. Res., 31 (4), 925-931.
(2) Brenan, K.E., Campbell, S.L., and Petzold, L.R., 1989. Numerical solution of initial-value problems in differential-algebraic equations, Elsevier Science Publishing Co., Inc., New York.
(3) Saad, Y., and Schultz, M.H., 1986. GMRES: A generalized minimal residual algorithm for solving nonsymmetric linear systems, SIAM J. Sci. Stat. Comp., 7, 856-869.
(4) Maier, R.S., Petzold, L.R., and Rath, W., 1994. Parallel solution of large-scale differential-algebraic systems, AHPCRC Preprint 94-014, University of Minnesota.
(5) Gear, C.W., 1971. Simultaneous numerical solution of differential/algebraic equations, IEEE Trans. on Circuit Theory, CT-18, No. 1, 89-95.
(6) Brown, P.N., Hindmarsh, A.C., and Petzold, L.R., 1993. Using Krylov methods in the solution of large-scale differential-algebraic systems, AHPCRC Preprint 93-033, University of Minnesota.
(7) Misra, D., 1994. Mixed finite element analysis of one-dimensional heterogeneous Darcy flow. Ph.D. dissertation, Submitted to the Graduate School of the University of Minnesota, Department of Biosystems and Agricultural Engineering, St. Paul, Minnesota.
(8) Letniowski, F.W., and Forsyth, P.A., 1991. A control volume finite element method for three-dimensional NAPL groundwater contamination, Int. J. Num. Methods in Fluids, 13, 955-970.

Mixed Finite Element Analysis of One-Dimensional Unsaturated Flow

D. Misra[1], J.L.Nieber[2], and H.V. Nguyen[2]. *[1]Army High Performance Computing Research Center, University of Minnesota. [2]Department of Biosystems and Agricultural Engineering, University of Minnesota.*

Introduction. Recent interests in unsaturated flow has shifted to the accurate simulation of the specific discharge or flux in the unsaturated soil because it was realized that this zone acted as a buffer for contaminants that eventually move to the ground water table. The presence of heterogeneities and sources/sinks in an unsaturated medium can significantly enhance or retard the migration of contaminants to the ground water table. In order to obtain accuracy in the simulation of contaminant transport it is essential to obtain highly accurate solutions of the Darcy flux in the flow domain. The mixed finite element method, originally introduced by (1), has found increased interest in linear steady state and transient groundwater flow problems due to its accuracy in simulating the Darcy flux (2,3). The basic idea of the method is to apply a finite element approximation to the conservation of mass and Darcy's law separately and obtain the values of the pressure and the velocity distributions by simultaneous solution of the two equations. The application of the mixed finite elements to unsaturated flow problems has been limited. The only applications that have been reported to the best of our knowledge is by (4, 5, 6, 7). There has been mixed response to the quality of solution obtained from application of the method.

Methodology. The name "mixed finite elements" originates from the fact that different orders of approximation (basis functions) are used to define the potential and the specific discharge in the finite element formulation. These orders of approximation are determined from the desired level of accuracy of the variable (e.g., specific discharge) that is important for the problem under consideration (7). For example, the choice of the interpolants or the basis functions for the zero-order mixed finite element method (8) is a constant interpolant for θ and ψ and a linear interpolant for q. Application of the method of weighted residuals and the Galerkin finite element method to the residual equations of the two governing equations of flow separately yields the global system of algebraic equations (7) that are solved simultaneously using the modified Picard method to obtain the desired solution for the pressure and the flux distribution. The results have been compared to solutions obtained using the conventional finite element scheme and the first order mixed finite element scheme (7).

Results and Discussion. A heterogeneous porous media with randomly assigned hydraulic properties was used as an example to study the effect of extremely heterogeneous conditions on the results of the conventional and the mixed finite element methods on a steady state unsaturated flow problem. The depth of the porous media was 40 m. The total depth was divided into 4 heterogeneous layers of 10 m thickness each. A point source of strength equal to 1.0 m^3/min was introduced at z = 30m. The boundary conditions for the problem are water table condition at z=0m and a constant flux of 0.001 m/min at z=40m. The hydraulic properties of each layer was described using the van Genuchten equations. For a non-heterogeneous medium these properties are given by: K_s = 0.04 m/min, θ_s = 0.35, θ_r = 0.0, α = 0.5 m^{-1} and v_n = 3.0. The heterogeneity of each layer was assigned using the scaling theory proposed by (9) where the matric potential and the hydraulic conductivity are scaled to simulate a heterogeneous porous media. In addition, the effect of the various averaging of the conductivities as described by (10) and (7) on the solution obtained from the different finite element schemes have been analyzed. The results of the matric potential head distribution are shown in Fig.1 for the conventional finite element scheme, Fig.2 for the zero-order and Fig.3 for the first-order mixed finite element schemes.

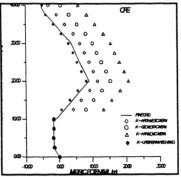

Fig. 1. Pressure head distributions obtained from the conventional finite element method.

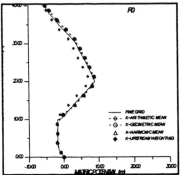

Fig. 2. Pressure head distributions obtained from the zero-order mixed finite element method.

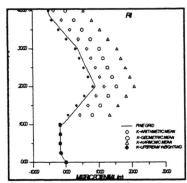

Fig. 3. Pressure head distributions obtained from the first-order mixed finite element method.

Conclusion. The matric potential and the flux distributions obtained from the zero-order mixed finite element scheme were more accurate than those of the other schemes for the example discussed above. The zero-order mixed finite element scheme yielded highly accurate flux distribution in the entire flow domain.

Literature Cited.
(1) Reissner, E. , 1950. On a variational theorem in elasticity, J. Math. Phys., 29, 90-95.
(2) Meissner, U. , 1973. A mixed finite element model for use in potential flow problems, Int. J. Numer. Methods Eng., 6, 467-473.
(3) Segol, G., Pinder, G.E., and Gray, W.G. , 1975. A Galerkin finite element technique for calculating the transient position of the salt water front, Water Resour. Res., 11(2), 343-347.
(4) Srinivas, C., Ramaswamy, B., and Wheeler, M.F. , 1992. Mixed finite element methods for flow through unsaturated porous media, In: Numerical Methods in Water Resources, edited by T.F. Russell, R.E. Ewing, C.A. Brebbia, W.G. Gray, and G.F. Pinder, Computational Methods in Water Resources, vol. 1, Computational Mechanics Publication and Elsevier Applied Science, USA, 239-246.
(5) Srivastava, R., and Yeh, T-C., J. , 1992. Comparison of mixed finite element method and a sequential solution approach for modeling flow and transport in a variably saturated porous media, Paper No. H42A09, Presented at the 1992 Fall meeting of the American Geophysical Union, Dec. 7-11, San Francisco, USA.
(6) Kolar, R.L. , 1992. Environmental conservation laws: Formulation, Numerical Solution, and Application, Ph.D. dissertation, Submitted to the Graduate School of the University of Notre Dame, Department of Civil Engineering and Geological Sciences, Notre Dame, Indiana.
(7) Misra, D., 1994. Mixed finite element analysis of one-dimensional heterogeneous Darcy flow. Ph.D. dissertation, Submitted to the Graduate School of the University of Minnesota, Department of Biosystems and Agricultural Engineering, St. Paul, Minnesota.
(8) Raviart, P.A., and Thomas, J.M. , 1977. A mixed finite element method for 2nd order elliptic problems, In: Mathematical Aspects of Finite Element Methods, Lecture Notes in Mathematics 606, I. Galligani and E. Magenes (eds.), Springer-Verlag, Berlin, 292-315.
(9) Miller, E.E., and Miller, R.D. , 1956. Physical theory for capillary flow phenomena, J. Appl. Phys., 27, 324-332.
(10) Haverkamp, R., and Vauclin, M. , 1979. A note on estimating finite difference interblock hydraulic conductivity values for transient unsaturated flow problems, Water Rsour. Res., 15 (1), 181-187.

A Numerical Study of the Impact of Transient-Hysteretic Flow on Nonreactive Solute Transport in the Vadose Zone

R.J. Mitchell[1] and A.S. Mayer[1]. *Department of Civil and Environmental Engineering and Department of Geological Engineering and Sciences, Michigan Technological University, Houghton, MI 49931, USA.*

Introduction.   Groundwater flow in the vadose zone is intermittent due to evaporation and redistribution periods between successive rainfall or irrigation events. Under intermittent flow conditions, hysteresis in the water content-pressure head relation can occur near the soil surface. There have been experimental (1) and numerical (2, 3) studies that have shown that hysteresis can retard solute transport in the vadose zone when compared to transient nonhysteretic and steady-state flow conditions. Inconsistencies in these studies arise due to variations in porous media types, flow regimes, number of irrigation and redistribution cycles, and evaporation considerations in the experiments. In the present study, numerical experiments are conducted with a fully hysteretic flow and transport model to monitor solute transport as a function of surface flux and porous media type. The experiments involve intermittent irrigation cycles with infiltration, redistribution, and evaporation stages. Comparisons are made between transient hysteretic and nonhysteretic cases and equivalent steady-state systems to further explore the use of steady-state parameters in the advection-dispersion equation (ADE) for describing solute transport driven by transient systems.

Materials and Methods.   A finite element flow and Eulerian-Lagrangian transport model was used for conducting one-dimensional, numerical experiments to investigate the effects of intermittent flow on solute transport in the vadose zone. Hysteresis was modeled using scaling relations of the hydraulic parameters and permeabilities (4, 5). The downward movement of a nonreactive solute slug was simulated for a series of 48-hour intermittent irrigation cycles. One cycle involved a 5-hour infiltration stage followed by a sequence of redistribution and evaporation stages totalling 19 and 24 hours, respectively. Two different infiltration rates were applied (q = 2.0E-04 and 5.0E-05 cm/s) along with a constant evaporation rate (q = -1.5E-06 cm/s). Five soils representing a range of soil classes were used in the numerical experiments. The van Genuchten draining and wetting hydraulic parameters for the five soils (Table 1) were obtained from fitted data subject to constrained conditions (6). Transient simulations were used to determine equivalent steady-state systems. Spatial moment analysis was used to monitor the position of the center of mass of the solute slug and to estimate a single time- and depth-averaged dispersivity and velocity used as steady-state transport parameters.

Table 1. van Genuchten parameters, where θ_{ds}, θ_{ws}, and θ_r are saturated draining and wetting and residual water contents (cm³/cm³), respectively, α_d and α_w are draining and wetting fitting parameters (cm¹), n is a fitting parameter, and K_s is the saturated hydraulic conductivity (cm/s).

Soil	θ_{ds}	θ_{ws}	θ_r	α_d	α_w	n	K_s
Rideau clay loam	0.416	0.416	0.288	0.0185	0.0473	2.637	2.66E-03
Ida silt loam	0.530	0.514	0.025	0.0158	0.0229	1.335	4.17E-04
Guelph loam	0.520	0.434	0.219	0.0098	0.0236	2.140	3.66E-04
Rubicon sandy loam	0.381	0.381	0.170	0.0136	0.0378	3.301	3.00E-04
Dune sand	0.301	0.301	0.101	0.0306	0.0527	6.779	5.83E-03

Results and Discussion.   The deviations between the position of the solute slug center of mass under transient hysteretic and nonhysteretic conditions varied greatly with soil type. The soils with large θ_{ds}/θ_{ws} (i.e., significant air entrapment properties) and α_w/α_d ratios exhibited the greatest deviations. Simulations using

95

the Dune sand and Rubicon clay loam showed insignificant differences between the respective solute center of mass positions under hysteretic and nonhysteretic conditions. Using a lower infiltration rate and an equivalent amount of water input did not significantly increase the center of mass deviations. Simulation results generated by seven irrigation cycles using the higher infiltration rate are shown in Fig. 1. Also indicated in Fig.1 is the center of mass position generated by equivalent steady-state simulations. In all cases, the velocity attained through moment analysis overestimated the solute slug position relative to the transient hysteretic and nonhysteretic cases. The results of this study indicate that hysteresis can retard solute transport in the vadose zone, but the degree of retardation is specific to soil type. In addition, the use of equivalent steady-state parameters in the ADE may over-predict the position of a solute slug under intermittent flow conditions.

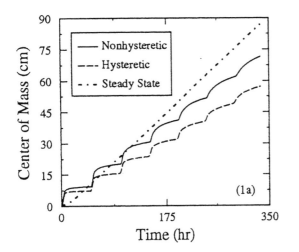 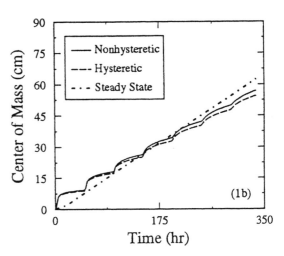

Fig. 1. Solute slug center of mass as a function of time: Guelph loam (1a) and Ida silt loam (1b).

Literature Cited.
(1) Sharma, M.L., and M. Taniguchi. 1991. Movement of a Non-Reactive Solute Tracer During Steady and Intermittent Leaching. J. Hydrol. 128:323-334.
(2) Pickens, J.F., and R.W. Gillman. 1980. Finite Element Analysis of Solute Transport Under Hystereitic Unsaturated Flow conditions. Water Resour. Res. 16:1071-1078.
(3) Russo, D., W.A. Jury, and G.L. Butters. 1989. Numerical Analysis of Solute Transport During Transient Irrigation: The Effect of Hysteresis and Profile Heterogeneity. Water Resour. Res. 25:2109-2118.
(4) Parker, J.C., and R.J. Lenhard. 1987. A Model for Hysteretic Constitutive Relations Governing Multiphase Flow 1: Saturation-Pressure Relations. Water Resour. Res. 23:2187-2196.
(5) Lenhard, R.J., and J.C. Parker 1987. A Model for Hysteretic Constitutive Relations Govening Multiphase Flow 2: Permeability-Saturation Relations. Water Resour. Res. 23:2197-2206.
(6) Kool, J.B., and J.C. Parker. 1987. Development and Evaluation of Closed-Form Expressions for Hysteretic Soil Hydraulic Parameters. Water Resour. Res. 23: 105-114.

Evaluation of the Validity of 2-Point Spatial Correlation Analysis of Microgeometry using Photoluminescent Volumetric Imaging

C. Montemagno and S. Ita. *Environmental Physics Group, Environmental Research Division, Argonne National Laboratory, Argonne, IL, 60439, USA.*

Introduction. Microgeometrical properties of soils are an important influence on hydrogeological processes in the vadose zone. Fluid infiltration and solute transport depend on the soil structure, which is determined by the grain sizes, grain shapes, pore sizes, and pore shapes. Early microgeometrical studies sought to establish a relationship between a specific property, such as porosity or pore throat size, and the hydrogeological processes [1,2]. More recent work has incorporated spatial correlation functions to provide a more complete, statistical characterization of the microgeometrical parameters [3,4]. These studies have relied upon highly idealized geometries to theoretically relate the spatial correlation functions to the microgeometrical parameters. In addition, they have focused on image analysis of two-dimensional sections and the assumption of isotropy to apply the measured two-dimensional correlation functions to the three-dimensional microgeometrical parameters. In this work we use the recently developed Photoluminescent Volumetric Imaging (PVI) technique [5] to assess the validity of applying two-point spatial correlation functions gathered from two-dimensional sections to the true three-dimensional microgeometrical parameters.

Materials and Methods. PVI is an optical visualization technique that can non-destructively image porous media to a high resolution. The porous media is composed of two constituents: an optical quality fused-quartz matrix and an immiscible fluid with a matching refractive index. The fluid is doped with trace amounts of property-selective fluorophores to highlight the desired structural features. The dye is excited by a planar laser light sheet. The fluorescent emissions are spectrally filtered and then recorded as digital computer files using a charged-coupled device camera. A series of two-dimensional images is generated by translating the laser sheet through the sample. The series is then processed into a three-dimensional data set that allows for quantitative study of the microscale properties. Nine different samples were measured to investigate the relationship between the microgeometrical parameters and the two-point spatial correlation functions (Table 1). Each two-dimensional sliced measured 3 mm by 3 mm, for a total sample volume of 9 mm^3. Pixel resolution was 5.6 µm. Two different dyes were used to highlight either the pore space or the grain-to-pore interface. The dye images that provided the most detailed information were used in subsequent analysis.

The two-point spatial correlation function, $S_2(\vec{\lambda})$, was calculated for each two-dimensional slice of each sample from the following:

$$S_2(\vec{\lambda}) = \int_{-\infty}^{\infty} f(\vec{x})f(\vec{x}-\vec{\lambda})d\vec{x} = FT^{-1}[F(\vec{k}) * f^*(\vec{k})] = \frac{1}{2\pi}\int_{-\infty}^{\infty} F(\vec{k})F^*(\vec{k})\exp(-\vec{k} \tag{1}$$

where $f(\vec{x})$ is the characteristic function, $\vec{x}$ is the Cartesian location, $\vec{\lambda}$ is the distance between the points, and $F(\vec{k})$ the Fourier transform of the characteristic function, and FT^{-1} the inverse of the Fourier transform. The two-point spatial correlation functions were then averaged to obtain a characteristic curve for each sample. The microgeometrical parameters were calculated using the methods described in [4].

Table 1. Grain size diameter distribution of samples by weight percentage.

Test	125-180 µm	180-250 µm	250-355 µm	355-500 µm	500-710 µm
1	100 %	0 %	0 %	0 %	0 %
2	0 %	100 %	0 %	0 %	0 %
3	0 %	0 %	100 %	0 %	0 %
4	0 %	0 %	0 %	100 %	0 %
5	0 %	0 %	0 %	0 %	100 %
6	9 %	12.5 %	18 %	25 %	35.5 %
7	14 %	16.5 %	19.5 %	23 %	27 %
8	20 %	20 %	20 %	20 %	20 %
9	0.01 %	2 %	11 %	16 %	71 %

Results and Discussion. The spatial correlation results (Figure 1, Table 2) are in poor agreement with the known microgeometrical properties. The estimated grain diameters are within the actual distributions only for the largest size distributions. The remaining properties do not follow the expected trends, e.g. increasing porosity and decreasing surface area with increasing grain diameter. In addition, significant variations were observed in the two-point spatial correlation functions between adjacent slices for each test. These results, for known isotropic samples, suggest that the two-point spatial correlation analysis technique is not presently suitable for application to field and laboratory samples. We believe that the discrepancies may be related to insufficient information in the two-dimensional slices, despite being larger than the dimensions of the representative elemental volume, but this requires further examination.

Table 2. Microgeometrical parameters calculated from spatial correlation analysis.

Test	Porosity, φ	Surface Area, SA (m^{-1})	Mean Grain Diameter (µm)	Mean Pore Diameter (µm)
1	0.22	8400	250	105
2	0.40	15000	275	105
3	0.25	6800	480	150
4	0.34	5900	260	225
5	0.49	4600	530	420
6	0.40	5700	355	215
7	0.45	7000	325	260
8	0.33	8200	540	160
9	0.45	5000	540	365

Literature Cited.

(1)   Ehrlich, R., S.K. Kennedy, S.J. Crabtree, and R.L. Cannon. 1984. Petrographic image analysis I. Analysis of reservoir pore complexes, J. Sed. Pet., 54:1365-1378.

(2)   Koplik, J., C. Lin, and M. Vermette. 1984. Conductivity and permeability from microgeometry. J. Appl. Phys., 56:3127-3131.

(3)   Berryman, J.G. 1987. Relationship between specific surface area and spatial correlation functions for anisotropic porous media. J. Math. Phys., 28:244-245.

(4)   Blair, S.C., P.A. Berge, and J.G. Berryman. 1993. Two-point correlation functions to characterize microgeometry and estimate permeabilities of synthetic and natural sandstones. Lawrence Livermore National Report, UCRL-LR-114182, 31 p.

(5)   Montemagno, C.D. and W.G. Gray. 1995. Photoluminescent volumetric imaging: a technique for the exploration of multiphase flow and transport in porous media. GRL, 22n4:425-428.

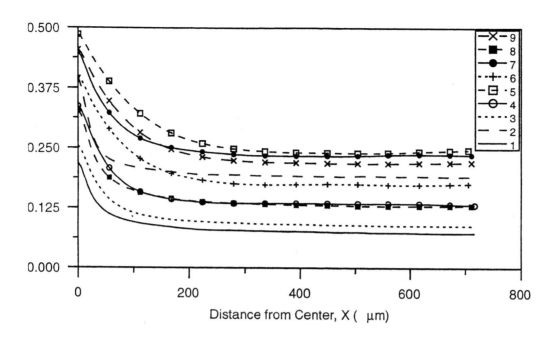

Figure 1. Average spatial correlation coefficients for all the tests.

Influence of Weaknesses on the Performance of a Capillary Barrier

H.J. Morel-Seytoux[1], and P. Meyer[2]. [1] *Hydrology Days Publications, 57 Selby Lane, Atherton, CA 94027-3926. [2]Dept. of Geosciences, Battelle Pacific Northwest Laboratories, Richland, WA 99352.*

Introduction. One of the crucial issues in the performance assessment of shallow land burial facilities for low-level waste (LLW) disposal is the analysis of water movement within the unsaturated soils above and around the waste. Water moving through the soil cover and into the disposal facility itself is a potentially significant means by which contaminants can be released to the environment. The primary objective of the infiltration analysis component of a LLW disposal facility performance assessment is to determine the amount of water coming into contact with the waste (1). The flux of water into the waste containment structure should be less than the natural recharge at the site while the site is operating as designed, and many years after construction when the facility may fail to operate as planned. The flow of water around and within a disposal facility will depend in large part on the local climate, hydrology, and geology and also on the specific facility design, construction, and operation. The purpose of this article is to illustrate one issue of concern in the analysis of infiltration and percolation foro the assessment of facility performance. The concern here is the possibility that for a good design, i.e., one that in theory should work, the presence of weaknesses in some locations might jeopardize the overall performance of the barrier.

Particular Physical Problem Considered. The system consists of two layers. The soil of the upper layer is fine (sand), which means that its "effective capillary drive" (2) is high, but its saturated hydraulic conductivity is low. The soil of the lower layer is coarse (gravel), which means that its attributes are, in relative terms, the opposite of those of the fine soil. The interface has a significant slope (typically a 5-to-1 slope) so that the problem is two-dimensional in nature in a vertical cross-section. A supply rate, r, is applied at the top of the sand layer (typically the water flux from an overlying clay layer). The same problem, but in one (vertical) dimension, was investigated previously (3). Of interest is the prediction of the evolution with time of the transmitted fraction across the interface. Eventually, as steady-state is approached, the interface saturates on the fine soil side. As more and more water piles over the interface the gravity drive overwhelms the capillary resistance of the fine soil and the entire incoming flux is transmitted. For the present problem configuration, on the contrary, water now moves laterally down the slope so that little ponded depth is expected at steady-state over the interface. In addition for a realistic and reasonably good design, at steady-state one expects that only a very small fraction of the incoming flux will be transmitted to the coarse layer - the gravel. The interface acts as a barrier to the flow in a permanent fashion. However, beyond a certain length the depth of water will have risen to such a level that the barrier is no longer effective. It is of interest to be able to predict that length for given soils and a design slope. A formula was previously derived (4) for that purpose. However, the solution assumed that while there is a sharp contrast between the two layers, each layer was perfectly homogeneous. Yet one would expect that the actual characteristics of the layers will vary from their projected design value. How will these deviations from the original design affect the performance of the system?

Illustration. The developed methodology is described elsewhere (4). Only a typical result is presented here. The data is the same "reference" data used in previous work (3) to facilitate comparisons. The saturated hydraulic conductivities and entry pressure heads (5) of the gravel and the sand are, respectively, 20 and 2 mm/h and 20 and 400 mm. In the case investigated, the vertical supply rate was 0.02 mm/h. The purpose of the run is to investigate the effect of a single weakness of limited extent on the performance of the barrier. Application of derived formulae for the data yields a value of 9 meters for the horizontal effective length of the barrier for a design critical level of fractional transmission of 2%. The weak spot is characterized by location at 5 meters, an extent of half a meter and a magnitude, i.e., reduction in capillary drive, of 28 cm. In the weak spot the capillary drive is only 62-28=34 cm. Naturally this is a significant reduction, the capillary drive value being only 55% that of the design value. The area under the peak of fractional transmission in the weak spot above the curve for the good design (see figure) is a measure of the reduction in performance of the barrier as a result of that particular weakness.

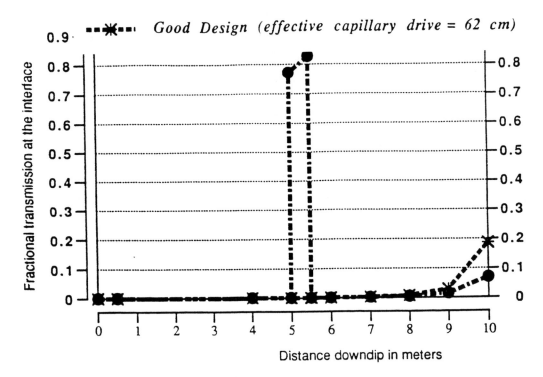

Fig. 1. Impact of a weakness in the sand on the performance of the capillary barrier as measured by distribution of the fractional transmission as a function of distance under steady-state.

Conclusions. Because there are many factors that affect the behavior of a barrier it is not possible to draw general conclusions from this limited investigation. What is clear is that it would not pay to cut corners. Provided is a basic tool that can help in the design of a capillary barrier by a thorough sensitivity analysis of the factors involved either in a deterministic framework or by complementing it with Monte Carlo simulations to assess the risk associated with a design.

Literature Cited.
(1) Meyer, P. 1993. Performance assessment of a hypothetical low-level waste facility: Application of an infiltration evaluation methodology. December 1993 Report #NUREG/CR-6114, PNL-8842, Vol. 1, Pacific Northwest Laboratory, Richland, WA 99352, 28 pp.
(2) Morel-Seytoux, H.J., and J. Khanji. 1974. Derivation of an equation of infiltration. Water Resour. Res. 10:795-800.
(3) Morel-Seytoux, H.J. 1992. The capillary barrier effect at the interface of two soil layers with some contrast in properties. HYDROWAR Report 92.4, Hydrology Days Publications, 57 Selby Lane, Atherton, CA 94027, 109 pp.
(4) Morel-Seytoux, H.J. 1994. Steady-state effectiveness of a capillary barrier on a sloping interface. Proc. 14th Annual AGU Hydrology Days, Hydrology Days Publications, 57 Selby Lane, Atherton, CA 94027, pp. 335-346.

Modeling Gravity Driven Fingered Flow in a Field Soil Having a Hydrophobic Layer

H. V. Nguyen[1], J. L. Nieber[1], C. Ritsema[2], L. Dekker[2] and D. Misra[3] . [1] *Department of Biosystems and Agricultural Engineering, University of Minnesota.* [2] *DLO Winand Staring Centre for Integrated Land, Soil and Water Research, Wageningen, The Netherlands.* [3] *Army High Performance Computing Research Center, University of Minnesota.*

Introduction. Gravity driven unstable flows in porous media are recognized to be an important process with respect to solute transport in the unsaturated zone (1). One mechanism for the initiation of unstable flow in porous media is a condition of hydrophobicity of the porous media (2). Evidence of fingered flow in hydrophobic field soils has been reported by (3). Recent continuous and nondestructive measurements of water content distribution in a Netherlands field soil containing a hydrophobic layer, reveals a very complicated wetting pattern with fingered flow being quite prevalent. The complicated pattern cannot be explained based on soil heterogeneity or soil macropores. Recent success at modeling gravity driven unstable flow (4) provides the potential to simulate the complicated patterns observed at field sites. In this presentation we attempt to simulate the patterns observed at the Netherlands site using a numerical solution intended for modeling gravity driven unstable flow.

Materials and Methods. A globally mass conservative finite element solution of the Richards equation is applied to the simulation of variably saturated flow in a rectangular region of homogeneous porous media. The solution employs rectangular elements with weighting of the internodal hydraulic conductivities. Capillary hysteresis in the water retention function for the porous medium is incorporated into the solution using an independent domain method (5). The water retention function is composed of a primary wetting function, a primary drainage function, and secondary wetting and drainage functions. With this set of retention functions it is possible to use the hysteresis model (5) to predict any wetting or drainage pathway.

Soil moisture and meteorological data have been collected from a field site near the village of Ouddorp in the Zeeland province of the Netherlands. Soil moisture data are collected at 7 depths (down to 0.7 meters) and 14 horizontal locations, at three hour intervals with an automated TDR data acquisition system. The width of the measurement domain is approximately 2.0 meters. The water table at the site resides at a depth of about 1.4 meters, and water table levels are acquired automatically with a pressure transducer placed in an observation well.

The soil profile is sandy texture composed of a 0.1 meter humic top layer '1' overlying a 0.30 meter hydrophobic layer '2', which is underlain by a hydrophilic layer '3' extending to the water table, see Figure 1. The boundary between the humic layer and the hydrophobic layer has a wavy form. Wetting and drainage retention functions have been measured for each of the soil horizons using the hanging water column method, and unsaturated hydraulic conductivity has been measured using the hot-air method.

A finite element grid using 28,000 grid points on a uniform vertical and horizontal spacing of 0.01 m is used to discretize the flow domain. The initial condition for soil water pressure in the domain was determined by interpolating the measured values of water saturation over the domain and computing the soil water pressure head from the interpolated saturations.

Results and Discussion. Distributions of water saturation in the soil profile at various times are shown in Figures 1, 2 and 3. Water saturations below 30% have been cut out of the diagrams to emphasize the preferential flow features. The distribution shown in Figure 1 is the initial condition and clearly illustrates the wavy form of the interface between the humic layer and the underlying hydrophobic layer. The water content in the hydrophobic layer is initially quite low, being less than 20%. The distribution in Figure 2 results following a rainfall of about 8 mm. It is observed that flow is penetrating preferentially at locations corresponding to the locations of the troughs in the humic layer-hydrophobic layer interface. This suggests that the form of the interface assists in the concentration of flow. The distribution shown in Figure 3 results after a rainfall of about 36 mm, and indicates further concentration of flows into the troughs of the humic layer-hydrophic layer interface. The fact that the concentrated flow does not diffuse outward, but instead moves downward as a growing perturbation is indicative of unstable flow. This unstable flow condition results from the hydrophobic nature of the second layer. Numerical simulation of this preferential flow process is still in progress.

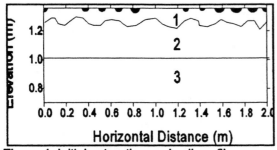

Figure 1. Initial saturation and soil profile.

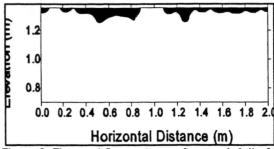

Figure 2. Fingered flow pattern after a rainfall of 8 mm.

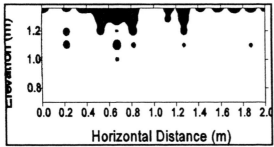

Figure 3. Fingred flow pattern after a rainfall of 36 mm.

Literature Cited.
(1) Glass, R.J., T.S. Steenhuis, and J.-Y. Parlange. 1989. Wetting front instability as a rapid and far-reaching hydrologic process in the vadose zone. J. Contam. Hydrol. 3:207-226.

(2) Raats, P.A.C., 1973. Unstable wetting fronts in uniform and non-uniform soils. Soil Sci. Soc. Am. Proc. 37:681-685.

(3) Ritsema, C.J. and L.W. Dekker. 1994. How water moves in a water repellent sandy soil. 2. Dynamics of fingered flow. Water Resour. Res. 30:2519-2531.

(4) Nieber, J.L., 1995. Modeling finger development and persistence in initially dry porous media, Geoderma, Accepted.

(5) Mualem, Y. 1974. A conceptual model of hysteresis, Water Resour. Res. 10:514-520.

Modeling Gravity Driven Fingered Flow in Unsaturated Porous Media

J.L. Nieber[1], H.V. Nguyen[1], and D. Misra[2]. [1]*Department of Biosystems and Agricultural Engineering, University of Minnesota.* [2]*Army High Performance Computing Research Center, University of Minnesota.*

Introduction. Gravity driven unstable flows in porous media are recognized to be an important process with respect to solute transport in the unsaturated zone (1) and migration of DNAPL's in the saturated zone (2). Numerous experimental investigations of the fingered flow process have been reported for laboratory conditions (3) and field conditions (4). Linear stability analyses have been the basis of a number of investigations of fingered flow theory (5), but only one attempt at simulation of fingered flow has been shown to be successful (6). Simulation of fingered flow is useful to increasing the understanding of fingered flow processes. In this presentation we show how the process of gravity driven fingering can be simulated, and demonstrate the influence of porous media properties on the stability of unsaturated flows.

Materials and Methods. A globally mass conservative finite element solution of the Richards equation is applied to the simulation of variably saturated flow in a rectangular region of homogeneous porous media. The solution employs rectangular elements with weighting of the internodal hydraulic conductivities. Capillary hysteresis in the water retention function for the porous medium is incorporated into the solution using an independent domain method (7). For this method all $\theta(h)$ pathways in the water retention function are completely defined by the function shown in Figure 1. The function consists of a primary wetting function, primary drainage function, secondary wetting function, and secondary drainage function. The initial conditions for a simulation are given by an air-dry medium except for a small distribution layer at the top of the rectangular domain. Small perturbations on the bottom of the distribution layer become the source for initiation of fingered flow. Water is supplied to the top of the domain at a constant rate less than the saturated hydraulic conductivity of the porous medium. Wetting and drainage sequences are simulated by turning the water supply on or off. Through this process we are able to simulate the growth of fingers as well as their persistence.

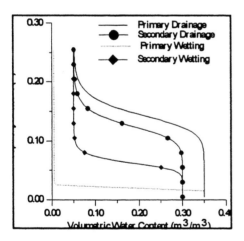

Figure 1. Water retention function.

Results and Discussion. It is found that small perturbations will grow into fingered flow when the water-entry capillary pressure on the primary wetting function is less than the air-entry pressure on the primary drainage function. Otherwise, the initial perturbation dissipates and the wetting front becomes stable. For the case where fingers grow it is found that: [1] The saturation distribution at and behind the wetting front have the same character as those observed in laboratory experiments (8); [2] The simulated finger width and finger velocity is in quite good agreement with those predicted by linear stability analysis; [3] When wetting and drainage sequences occur, initial fingered pathways persist as preferential flow paths for future wetting events; [4] The number of fingers, their width and velocity, are strongly dependent on the form of the initial perturbation. The distribution of saturation along the central axis of a finger is shown in Figure 2. The fingered flow pattern for the case of a regular initial perturbation pattern is shown in Figure 3.

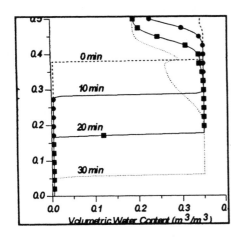

Figure 2. Water content distribution along the central axis of the center finger shown in Figure 3.

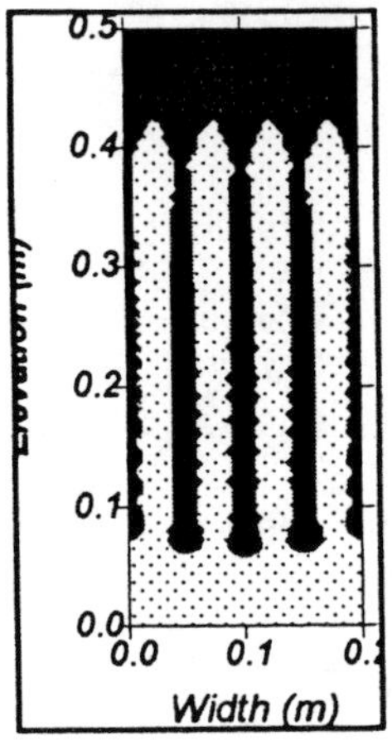

Figure 3. Fingered flow at 20 minues.

Literature Cited.
(1) Glass, R.J., T.S. Steenhuis, and J.-Y. Parlange. 1989. Wetting front instability as a rapid and far-reaching hydrologic process in the vadose zone. J. Contam. Hydrol. 3:207-226.
(2) Ritsema, C.J. and L.W. Dekker. 1994. How water moves in a water repellent sandy soil. 2. Dynamics of fingered flow. Water Resour. Res. 30:2519-2531.
(3) Glass, R.J., J.-Y. Parlange and T.S. Steenhuis. 1989. Wetting front instability, 2. Experimental determination of relationships between system parameters and two-dimensional unstable flow field behavior in initially dry porous media, Water Resour. Res. 25:1195-1207.
(4) Held, R.J. and T.H. Illangasekare. 1995. Fingering of dense nonaqueous phase liquids in porous media, 1, Experimental investigation. Water Resour. Res. 31:1213:1222.
(5) Glass, R.J., J.-Y. Parlange and T.S. Steenhuis. 1989. Wetting front instability, 1.Theoretical discussion and dimensional analysis, Water Resour. Res. 25:1187-1194.
(6) Nieber, J.L., 1995. Modeling finger development and persistence in initially dry porous media, Geoderma, Accepted.
(7) Mualem, Y. 1974. A conceptual model of hysteresis, Water Resour. Res. 10:514-520.
(8) Liu, Y., T.S. Steenhuis and J.-Yves Parlange, 1994. Formation and persistence of fingered flow in coarse grained soils under different moisture contents, J. Hydrol., 159:187-195.

Use of the Cosolvency Model to Predict Sorption Kinetics and Equilibria of Diuron and Naphthalene on Organoclays

P. Nkedi-Kizza[1] and V. Nzengung[2]. [1]University of Florida, FL 32611, USA. [2]University of Georgia, Athens, GA 30602, USA.

Introduction. Since the pioneering research of Nielsen and Biggar (1) and Biggar and Nielsen (2) on solute transport in soil materials, one major process that has limited our ability to correctly model solute transport in soils is sorption kinetics and the related sorption equilibria of organic pollutants. Due to the concern of pollution of groundwater from organic chemicals, there is currently interest in using clays treated with alkylammonium cations (organoclays) to increase the sorption of hydrophobic organic chemicals by landfill liners, clay barriers and water treatment systems (3). Our suspicion that at waste disposal sites a mixture of water and organic solvents might be found (which could influence sorption kinetics and equilibria of organic pollutants) lead us to use the solvophobic theory to predict sorption of two probe organic compounds (diuron and naphthalene) on two organoclays (TMPA- and HDTMA-clay) in varying mixtures of water and methanol.

Materials and Methods. The method of Nzengung (3) was used to transform the Na-montmorillonite into benzyldimethyltetradecylammonium clay (BDTDA-clay) or trimethylphenylammonium clay (TPMPA-clay). Naphthalene and diuron (3-3(4-dichlorophenyl)-1,1-dimethylurea), with similar hydrocarbonaceous surface area (HSA) were used as ^{14}C and ^{12}C in batch reactors to determine kinetic parameters (k_d and k_a) and the sorption equilibrium parameter (K_p) on TMPA and HDTDA-clay (4). The kinetic data were simulated with the One Site Kinetic Model (OSKM), (5) and obtained parameters were used in the cosolvency model to predict kinetic parameters in aqueous systems. The equilibrium sorption coefficients (K_p) in mixed solvents were also simulated with the cosolvency model (4). The cosolvency model for sorption equilibria is given by:

$$\ln(K^m_p) = \ln(K^w_p) - \alpha\sigma f_c \qquad [1]$$

The OSKM in a batch reactor is conceptualized as (3,5):

$$C \rightleftharpoons S \qquad [2a]$$
$$K^m_p = (k^m_a/k^m_d)/\theta^m \qquad [2b]$$
$$\ln(k^m_d) = \ln(k^w_d) + A\alpha\sigma f_c \qquad [3]$$
$$\ln(k^m_a/\theta^m) = \ln(k^a_w/\theta^w) - (1-A)\alpha\sigma f_c \qquad [4]$$
$$k_d/\theta = k_a. \qquad [5]$$

Where C is the solution concentration (g/ml), S is the sorbed concentration (g/g), k_a and k_d are adsorption and desorption rate coefficients, respectively; K_p is the sorption coefficient (ml/g), θ is the soil:solution ratio, α is the constant for solvent sorbent interactions, σ is the cosolvency power, f_c is the volume fraction of methanol, w and m denote aqueous and cosolvent systems, respectively.

Results and Discussion.

The K_p-f_c Relationship. In Fig.1, the relative sorption coefficient (K^m_p/K^w_p) vs f_c for naphthalene and diuron sorption on both organoclays are shown. The pooled data can be described by a single line as expected from Eq. 1, since both diuron and naphthalene have essentially the same σ value. The cosolvency model describes all data fairly well, similar to soil data reported for diuron in methanol as a co-solvent (4).

The (k_a or k_d) -f_c Relationship. The log-linear relationships predicted by Eqs. 3 and 4 for k_a and k_d, respectively, describe the data reasonably well (Fig. 2). The scatter in the data between solutes and organoclays might be an indication of slight differences in α since σ is essentially the same for both solutes. It appears the model is more sensitive to k_d. All data presented indicate that the cosolvency model can be used to predict kinetic and equilirium parameters for sorption of these two probe compounds on the two organoclays in mixed solvents.

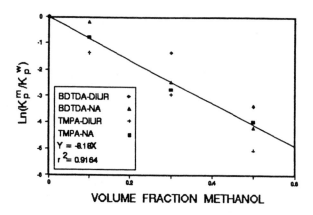

Fig. 1. Relative sorption coefficient (K^m_p / K^w_p) as a function of volume fraction cosolvent (f_c).

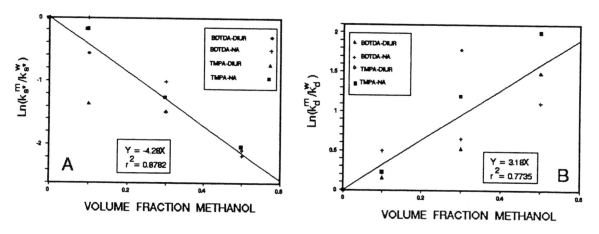

Fig. 2. Relative adsorption rate coefficient (k^m_a / k^w_a), and relative desorption rate coefficient (k^m_d / k^w_d) as a function of volume fraction cosolvent (f_c).

Literature Cited.

(1) Nielsen, D.R., and J.W. Biggar. 1961. Miscible Displacement in Soils: I. Experimental information. Soil Sci. Soc. Amer. Proc. 25:1-5.

(2) Biggar, J.W., and D.R. Nielsen. 1963. Miscible Displacement: V. Exchange Processes. Soil Sci. Soc. Amer. Proc. 27:623-627.

(3) Nzengung, V.A. 1993. Organoclays as Sorbents for Organic Contaminants in Aqueous and Mixed-Solvent Systems. Ph.D. Dissertation, Earth and Atmospheric Sciences, Georgia Institute of Technology, Atlanta, Georgia.

(4) Nkedi-Kizza, P., P.S.C. Rao, and A.G. Hornsby. 1985. Influence of Organic Cosolvents on Sorption of Hydrophobic Organic Chemicals by Soils. Environ. Sci. Technol. 19:975-979.

(5) Selim, H.M., J.M. Davidson, and R.S. Mansell. 1976. Evaluation of a two-site adsorption-desorption model for describing solute transport in soils. Proc. Summer Computer Simulation Conf., Washington D.C.

Improvement of Soil Particle Size Analysis Through Gamma-Ray Attenuation

J.C.M. de Oliveira[1], C.M.P. Vaz[2], K. Reichardt[1,3], O.O.S. Bacchi[1]. [1] *Center for Nuclear Energy in Agriculture, CENA, Av. Centenário 303, 13400-970 Piracicaba, SP, Brazil.* [2] *National Center for Research and Development of Agricultural Instrumentation, CNPDIA/EMBRAPA, Rua XV de Novembro 1452, 13560-970 São Carlos, SP, Brazil.* [3] *Department of Physics and Meteorology, ESALQ, Av. Pádua Dias 11, 13418-900 Piracicaba, SP, Brazil.*

Introduction. Particle size distribution of soils is a very important issue in describing soils as porous materials. Its relation with most fluid retention and transmission properties of the soil makes particle size distribution the most common physical property used to characterize a soil. Among the several methods suggested for the measurement of soil particle sizes, the one proposed by (1) presents several advantages over the two most commonly used: the pipette and hydrometer methods (2). The advantage of not interfering with the sedimentation process gives to the gamma method a much higher credibility and permits a more generalized use of Stokes' sedimentation law. It will be shown here that with the aid of a system to monitor the movement of the sedimentation flask, the measurement time may be reduced about ten fold, the possibility for automation becomes very practical, making the method more attractive than the pipette and hydrometer methods for routine analysis.

Materials and Methods. The method is based on the attenuation of a gamma-ray beam by a soil particle suspension prepared in the classical way (2), as described by (1). The equation that describes the attenuation process of a gamma-ray beam by a heterogeneous system under sedimentation, composed of an acrylic container, soil particles, water and sodium hydroxide (used for soil aggregate dispersion) is

$$I = I' \exp(-\mu_p \rho_p X_p - \mu_w \rho_w X_w - \mu_h \rho_h X_h - \mu_c \rho_c X_c) \tag{1}$$

where I and I' are gamma ray beam intensities; X the thickness of the absorbing materials; ρ their specific masses and μ their mass attenuation coefficients; the subscripts p, w, h, and c indicate, respectively, soil particles, water, sodium hydroxide and container. The actual interspersing of particles, water, and hydroxide has been replaced by an equivalent sequence of length X_p of specific mass ρ_p, X_w of specific mass ρ_w and length X_h of specific mass ρ_h. Neglecting X_h and making a measurement without soil (I_o), equation 1 becomes

$$I = I_o \exp(\mu_w \rho_w X_p - \mu_p \rho_p X_p) \tag{2}$$

which is the attenuation equation valid during the sedimentation process. According to Stokes's law, during the sedimentation process, after a time t measured from the start of the sedimentation process, there is a depth h measured from solution surface, at which only particles of diameter less than d are found in suspension. Under isothermic conditions, the time t is related to h according to

$$t/h = 18\eta / [g\, d^2 (\rho_p - \rho_w)] \tag{3}$$

where η is the viscosity of the liquid and g the acceleration of gravity. Within the gamma-ray beam of cross-sectional area A, passing through the suspension at depth h, the volume of suspension within the beam is AX_i, X_i being the internal container dimension. The soil-particle volume within the beam at a given time is AX_p, with mass $\rho_p A X_p$. The mass of particles per unit volume of suspension is thus $\rho_p X_p / X_i$, in g.cm^{-3}. Converting this to concentration C in g.l^{-1}, we have $X_p = C X_i . 10^{-3} / \rho_p$, which when substituted into equation 2 and solved for C gives

$$C = [10^3 \ln(I_o / I)] / [X_i (\mu_p - \mu_w \rho_w / \rho_p)] \tag{4}$$

Most methods (e.g. pipette, densimeter, gamma-attenuation) suggest that the depth of sampling be fixed and sediment concentration C be measured as a function of time. Since the gamma method easily permitts the change of the position of the beam pathway, we suggest moving the sample vertically during sedimentation, and measure C at different combinations of h and t. Equation 3 indicates that particles of the same diameter d can be found at different combinations of t and h. Therefore, making attenuation measurements at decreasing values of h, which is achieved by moving the sample upwards, at choosen Δh increments, starting from the bottom of the sedimentation container, the total scanning time can be largely reduced. On the other hand, measuring at increasing values of h (moving the sample downwards), particles of the same diameter can be found at different times, and measurements of a given C can be repeated.

Results and Discussion. Figure 1 illustrates the particle size distributions of three samples. The total scanning times to obtain complete particle size distributions are on the order of 20 minutes. When comparing the time with that spent performing pipette or densimeter measurements, it is clear that the reduction is substantial. This point is important in routine particle size analysis (PSA), and adding the automation advantages, the methodology here recommended becomes a great improvement of the methodology originally presented by (1). The coefficients of variation (CV) of the concentrations obtained in three consecutive sedimentation runs were on the order of 2.5% in the sand/silt range, and of 8.5% in the silt/clay range. This increase in CV is due to the relatively low concentrations when the suspension contains mostly clay particles. This problem could be minimized by increasing Δt for large t, when the sedimentation process has considerably slowed down.

Conclusion. The innovation here presented of moving the sample vertically in the method of PSA of (1) turns it into a more rapid one (about 20 minutes for each sample), continues with the advantage of no perturbation of the sedimentation process and generalizes the method for other sedimentation studies. It permits the measurement of any particle size fraction and a greater detail of the sand fractions. With all these attractives it presents a high competitiveness in relation to the traditional routine PSA methods.

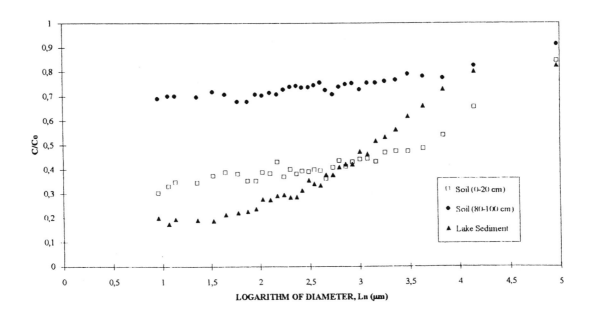

Fig. 1. Examples of particle size distributions obtained by gamma attenuation method.

Literature Cited.

(1) Vaz, C.M.P., J.C.M. de Oliveira, K. Reichardt, S. Crestana, P.E. Cruvinel, and O.O.S. Bacchi. 1992. Soil mechanical analysis through gamma-ray attenuation. Soil Technology 5: 319-325.
(2) Gee, G.W., and J.W. Bauder. 1986. Particle size analysis. In: A. Klute (ed.) Methods of soil analysis. Part I, 2nd ed., Agron. Monogr. 9. ASA and SSSA, Madison, WI, 383-411.

Are Existing Formulations for Evaporation From Bare Soil Useful?

M. Parlange[1], A. T. Cahill[1], J. Albertson[1], M. Mata[1], B. Moore[1], T. Jackson[2], P. O'Neill[3]. *[1]Hydrologic Science, University of California, Davis, CA 95616, USA. [2] Hydrology Lab., USDA, Beltsville, MD 20705, USA. [3] NASA Goddard, Greenbelt, MD 20771, USA.*

Introduction. There has been much interest on how evaporation from bare soil surfaces should be formulated in practice (1,2,3,4,5). Most evaporation methods rely on empirical functions based on near surface volumetric soil moisture (θ). However, there have been relatively few simple field tests of the existing models. Ultimately, in practical calculations, the instrument for the soil moisture measurement (e.g. passive microwave platform, neutron probe, time domain reflectometry, etc.) will dictate the appropriate evaporation model formulation. Critical issues that need to be addressed include: over what depth the sensor measures θ, at what time interval θ known and what is the soil-atmosphere interface moisture content inferred from the measurement. Field studies are carried out at the UC Davis Campbell Tract to investigate the appropriateness of S and L band passive microwave θ measurement, as well as in situ techniques of time domain reflectometry (tdr) and neutron probe in bare soil evaporation models.

Materials and Methods. Most bare soil evaporation models are written as the α method,

$$E = \rho C_E u(\alpha q^*(T_s) - q) \tag{1}$$

and the β method,

$$E = \rho C_E u\beta(q^*(T_s) - q) \tag{2}$$

where E is the evaporation rate, ρ is the air density, u is the wind speed, C_E is the bulk coefficient of evaporation (formulated in the context of mean Monin-Obukhov similarity theory), q the specific humidity of the air, $q^*(T_s)$ the saturation value at the soil surface temperature T_s, and α and β are coefficients which represent the soil surface moisture availability. The moisture availability parameters were derived from 20 minute time averaged measurements of E, q, u and T_s. For the preliminary analysis presented here θ was simultaneously measured over the 20 minute periods using TDR placed 2 cm below the soil surface.

Results and Discussion. Preliminary results for α and β [see Figures 1 and 2] as functions of θ from the tdr measurements clearly demonstrate the error that will arise in flux calculations. This is interesting given that most atmospheric numerical models (1,2,3,4,5) currently rely on the plotted lines in Figures 1 and 2. Extensions for the additional soil water sensors are presented at the conference along with appropriate functions for flux parameterization.

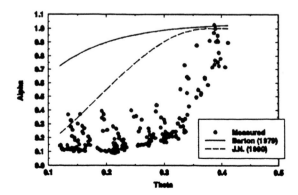

Figure 1. α versus θ; the behavior of α is not represented by the existing models.

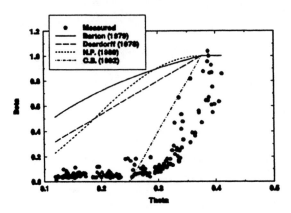

Figure 2. β versus θ; the Crago-Brutsaert model behaves the best, again most models for meteorological applications fail to properly partition the available surface energy.

Literature Cited.
(1) Barton, I.J. 1979. A parameterization of the evaporation from nonsaturated surfaces. J. Appl. Meteor. 18:43-47.
(2) Jacquemin, B. and J. Noilhan. 1990. Sensitivity study and validation of land surface parameterization using the HAPEX-MOBILHY data set. Bound.-Layer Meteor. 52:93-134.
(3) Deardorff, J.W. 1978. Efficient prediction of ground surface temperature and moisture, with inclusion of a layer of vegetation. J. Geophys. Res. 83:1,889-1,903.
(4) Noilhan, J. and S. Planton. 1989. A simple parameterization of land surface processes for meteorological models. Mon. Wea. Rev. 117:536-549.
(5) Crago, R.D. and W. Brutsaert. 1992. A comparison of several evaporation equations. Water Resour. Res. 28:951-954.

Soil Water Dynamics of a Water Regulated Raised Bog

G. Persson. *Swedish University of Agricultural Sciences, Department of Soil Sciences, P.O. Box 7014, S-750 07 Uppsala, Sweden.*

Introduction. A peat bog represents a hydrological system with a shallow groundwater which implies a sensitive relationship between the water table and the water content in the unsaturated zone. When forest plantations are made at such virgin soils with a low pH-value, problems arise regarding root establishment . A shallow root zone gives a sensitive water supplying system despite the shallow groundwater table. The water dynamics of a regulated raised bog were determined for four years and the relationship between groundwater level and soil water content was investigated.

Materials and methods. A water balance study at a raised bog was performed during 1980-1983. The bog was drained by ditches at about 20m distance. Runoff measurements were made in one of the main ditches representing an area of about 12 ha. In one of the studied years groundwater levels and soil water content were measured at this subarea once a week. Groundwater tubes were placed in lines across the ditches and capacitance sensors were placed in the vicinity of the tubes. Soil probes were taken with a peat auger down to groundwater level about once a month and the probes were cut in 10 cm long samples, for which the water content were determined gravimetrically. The capacitance sensors were calibrated against the soil probes determinations.

A physically based soil model was used to simulate the water dynamics during four growing seasons. The model was calibrated against runoff measurements and groundwater levels for one of the seasons. The climatic data used as input to the model were daily averages of air temperature, air humidity and windspeed together with daily sums of precipitation and radiation. The soil physical conditions described as the pF-curve and the saturated hydraulic conductivity were taken from a similar sphagnum peat soil.

Results. The water content at 15 cm depth was well reflecting the dynamics of the groundwater level with the lowest water content and groundwater level at the end of August and the highest values after rains in September (Fig.1.). The groundwater levels, expressed as a distance from the soil surface, is most shallow closest to the ditches, compared to the groundwater levels in between the ditches (Fig.2.). The highest difference, found during the driest period, is about 15 cm. The mean groundwater level is between 30 cm and 70 cm depth below soil surface. The runoff regime varies between years with sometimes high spring flows and zero flow in midseason and in some years a more even flow through the season (Fig.3.)

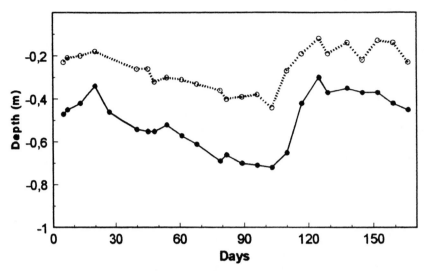

Fig.1. The man groundwater level during the growing season together with the mean water content (vol-%) at 15 cm depth determined by capacitance sensors.

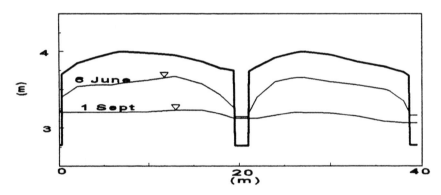

Fig.2. Cross-section between ditches with groundwater levels at two days.

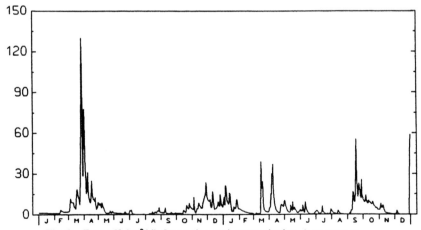

Fig.3. Runoff (m³/d) from the subarea during two years.

Using Deforming Finite Element Models to Create a Regional Scale Saturated-Unsaturated Flow and Transport Model

D. Purkey[1], and W. Wallender[2]. [1] *Hydrologic Science, University of California, Davis, CA 95616, USA.* [2] *Hydrologic Science and Biological and Agricultural Engineering, University of California, Davis, CA 95616, USA.*

Introduction. In recent years, a series of saturated-unsaturated flow and transport models have been developed (1,2). These models describe the movement of water and solutes in response to a continuum of hydraulic head and concentration values in a region of porous media. These saturated-unsaturated models treat any free surface in the system as part of the continuum, as indeed it is in nature. This approach differs from the inclination of earlier investigators to treat any free surface as a boundary between separate saturated (groundwater) and unsaturated (vadose) systems.

On a philosophical level, one cannot argue that the old reductionist view is a better representation of nature. Nonetheless, on a practical level, saturated-unsaturated models may not be appropriate in all cases, particularly for regional, or km, scale investigations. A review of the literature finds most have, to date, been applied on a cm or m scale (e.g. 1,2). Thus, for example, as a tool for assessing management options on the scale of a water district, it seems, the reductionist approach may still have some utility. This poster describes an ongoing investigation of the use of separate deforming finite element models of the groundwater and vadose zone systems in a coupled porous media flow and transport model.

Modeling Approach. The notion of coupling models of saturated and unsaturated flow is not entirely new. One-dimensional Richards Equations have been coupled with both one- (3) and two-dimensional (4) Boussinesq equations. What is unique about the current investigation is the attempt to describe the transient evolution of the saturated zone, via temporal changes in the position of the free surface, without invoking the Dupuit Assumptions. This is achieved by replacing the Boussinesq Equation with an algorithm developed to describe, non-steady, fully two-dimensional flow in the saturated zone using a deforming finite element mesh (5). Here, temporal fluctuations of the water table are mimicked by the deformation of the elements which make up the free surface. This deformation, in turn, results in the deformation of the one-dimensional Richards Equation models which describe the vadose zone as they shorten and lengthen in response.

Results. The saturated portion of the coupled model has been successfully applied to the problem of the growth and decay of a groundwater mound. An analytical solution is available for this problem when the porous media is homogeneous and aquifer recharge is temporally and spatially uniform (6). Results for the analytical and numerical approaches compare favorably (Figs. 1 and 2). When applied to two simple heterogeneous geometries the numerical approach gives reasonable results (Fig. 3). In all cases, the development of vertical gradients in the saturated region, which are ignore when using the Boussinesq Equation, are described by the deforming finite element model.

Conclusion. Further refinement of the coupled saturated-unsaturated flow and transport model is underway. The model will be applied a 16 km long transect across the Panoche Irrigation District, California in order to investigate how changes in management patterns on the landscape translate to changes in regional-scale flow and transport processes.

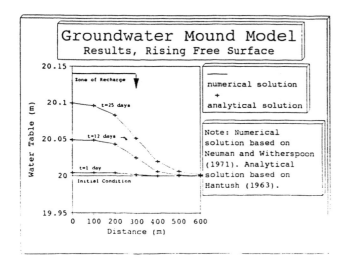

Fig. 1.

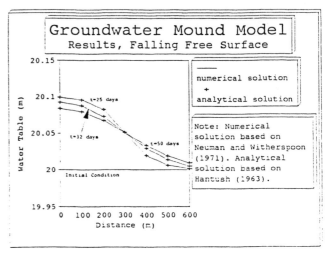

Fig. 2.

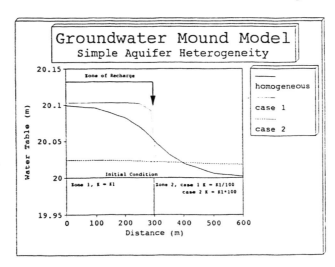

Fig. 3.

Literature Cited.

(1) Kool, J.B. and M.Th. van Genuchten. 1991. HYDRUS one-dimensional variably saturated flow and transport model including hysteresis and root water uptake. Version 3.31. Res. Report No. 124, US Salinity Lab., USDA, ARS, Riverside, CA.

(2) Simunek, J., T. Vogel and M.Th. van Genuchten. 1994. The SWMS_2D code for simulating water flow and solute transport in two-dimensional variably saturated porous media. Version 1.2. Res. Report No. 132, US Salinity Lab., USDA, ARS, Riverside, CA.

(3) Pikul, M.F., R.L. Street, and I. Remson. 1974. A numerical model based on coupled one-dimensional Richards and Boussinesq Equations. Water Resources Research. 10:295-302.

(4) Abbott, M.B., J.C. Bathhurst, J.A. Cunge, P.E. O'Connell, and J. Rasmussen, 1986. An introduction to the European Hydrologic System, 2. Structure of a physically-based, distributed modelling system. Journal of Hydrology. 87:61-77.

(5) Neuman, S.P. and P.A. Witherspoon. 1971. Analysis of nonsteady flow with at free surface using the finite element method. Water Resources Research. 7:611-623.

(6) Hantush, M.S. 1963. Growth of a groundwater ridge in response to deep percolation. Symposium on Transient Groundwater Hydraulics. Ft. Collins, CO.

Daily Rainfall Variability at a Local Scale (1,000 ha) in Piracicaba, SP, Brazil and its Implications on Soil Water Recharge

K. Reichardt[1,2], L. R. Angelocci[1], O. O. S. Bacchi[2], J. E. Pilotto[2]. [1]*Department of Physics and Meteorology, ESALQ, Av. Pádua Dias 11, 13418-900 Piracicaba, SP, Brazil.* [2]*Center for Nuclear Energy in Agriculture, CENA, Av. Centenário 303, 13400-970 Piracicaba, SP, Brazil.*

Introduction. It is well known that rainfall events in the tropics can have great variability over short distances. Commonly one observes that it is raining at a place and very close to it there is no rain. In many circumstances it is possible to observe on the ground the moving limit of the wetted front. These facts are extremely important in the planning of rainfall measurements in space and time. This variability has great implications in the establishment of weather networks. Networks usually represent a collection of point data, the accuracy of each depending on the sensor characteristics, calibration, and proper exposure. Assuming as minimal the instrumental errors at each obsevation point, the selection of representative points is a difficult task. In Brazil rain gauge networks started only at the end of the last century, and up to date, in the most advanced part of the country, they represent a collection of points, each belonging to one county. As an example, in the network of the State of São Paulo, the county of Piracicaba with an area of 142,100 ha, is represented by only one observation point. How well does this point represent the county ? The present study considers this variability on a smaller space scale, here called "local", which represents an area of the order of 1,000 ha, which is about the size of the Campus of the University of São Paulo at Piracicaba. Since it is common to rain in one part of the campus and in another not, it is important to characterize this variability. The most frequently weather variable referred to in the literature is rainfall. Several authors [(1), (2), (3)] present network designs of special interest for this study. Among the reports on rainfall spatial variability, one important (4) is a study of several weather variables including rainfall, using data of 24 stations of a medium scale network in USA. He concludes that to monitor precipitation at a level to explain > 90% of the variation between sites, the spacing between them should be less than 5 km. With respect to the effect of rainfall variability on soil water changes and on consequent yield, on a local basis, no reports have been found. The effect of soil variability in estimating evapotranspiration from the water balance equation (5) was performed in a study using a strip of 5×125 m^2, considering the rainfall constant, although the rain gauge was located about 200 m from the center of their plot. Which would be the minimum size of an experimental area in order to be allowed to consider precipitation constant ? This study represents an effort to answer these kind of questions.

Materials and Methods. Nine rainfall collection points were choosen as randomly as possible within the area of the Campus of the University of São Paulo in the county of Piracicaba, Brazil (22° 44' S; 43° 33' W), at an altitude of 580 m above sea level, 250 km inside the continent. The tenth collection point, considered as a standard, is the official weather station of the campus, for which 78 years of rainfall data are available. The distribution of the collection points cover an area of about 1,000 ha. The closest and the farest points from the standard are, respectively, 990 and 2,550 m apart. The raingauges choosen have a collection area of 300 cm^2, installed 1.5 m above ground level, being free of any obstacle in a circle of at least 20 m radius. Their accuracy of measurement is 0.1 mm. Data collection started November 1993 and ended October 1994, completing one full year.

Results and Discussion. For the kind of study here focused it is important to the reader to observe raw data, in order to obtain a feeling of the variability that can be expected. Table 1 presents selected raw data, for all ten observation points, together with their means, standard deviations, and coefficients of variation (C.V.). Daily C.Vs. varied from 2.2% to 169.3%. From the 87 rainy days of the year , 15 present a C.V. below 10% and 8 above 100 % . For the days with C.V. below 10%, precipitations were indistinctly of low and high intensities, however for the days with C.V. above 100%, precipitations were all of very low intensity, most of them including several zeros, i.e. with several observations very close to the accuracy of the rain gauges. During the collection period it was also observed that the magnitude of the variability was independent of the type of rain, i.e. whether it is of convective, front or other origin. In order to calculate the cumulative precipitation for the whole observation period, missing data were replaced by the respective daily averages. (4) replaced missing data by weighted averages using five nearest stations and the inverse separation distance law. Since the data of this study did not present any correlation with distance, simple averages were maximum values are, respectively, 1,020 and 1,141 mm, and the coefficient of variation is 3.06. To quantify the spatial variability (5) and (6) plotted the correlation coefficient between data of pairs of obsevation points, with respective separation distances. This approach was also here used, and no correlation with distance was found. This fact indicates that, although there is a great variability among observation sites, this variability has a random characteristic. The fact that a more dense cloud is positioned over some specific rain gauges is a compleately randomic process, at the studied scale. Therefore it is not possible to estimate at which distance a given observation explains a given fraction of the respective observation at the standard obsevation point, as (4) did. The results however show the importance of the variabilty in space. Authors avoided to take montly and quarterly averages because the number of rainfall days is very variable, and because the available data corresponds only to one year of observation. Data, however, clearly indicate that rainfall data collected at the standard observation point do not represent areas as far as 1,000 to 2,500 m apart, for daily, monthly or even quarterly averages, which is the case of annual crops. For yearly totals the coefficient of variation was 3.06%, indicating that any of the observation points can replace the standad observation point.

Table 1. Selected days of rainfall data (mm) for all ten observation points, with averages, standard deviations and coefficients of variation.

Date	P1	P2	P3	P4	P5	P6	P7	P8	P9	PStd	Ave	S.D.	C.V.
12/28	27.8	27.2	28.4	29.5	24.3	28.4	27.5	31.8	29.6	28.9	28.34	1.94	6.8
01/31	00.0	0	0.7	0	0.2	0.3	0	2.1	0	1.6	0.49	0.76	154.9
02/19	18.8	22.0	0.5	13.2	22.3	6.8	17.7	15.1	17.6	12.3	14.63	6.81	46.5
04/21	53.1	48.0	49.5	56.0	55.8	50.6	56.6	52.3	50.5	49.4	52.18	3.09	5.9
04/26	15.5	8.8	29.9	12.0	3.5	16.6	6.4	30.2	16.8	24.5	16.42	9.33	56.8
05/22	0.0	0.2	0.7	0.5	0.1	0.5	0.8	0.5	0.2	0.6	0.41	0.27	65.5
06/22	14.9	15.0	14.3	14.8	14.8	13.7	13.9	14.0	14.7	14.3	14.44	0.46	3.2
Tot:	1029.0	1043.1	1103.4	1076.0	1087.4	1109.8	1073.7	1141.1	1069.7	1106.1	1083.9	33.2	3.1

Literature Cited.

(1) Eagleson, P.S. 1967. Optimum density of rainfall networks. Water Resour. Res., 3: 1021-1033.

(2) Morin, G., J. Fortin, W. Sochanska, and D.A. Wilhite. 1979. Use of principal component analysis to identify homogeneous precipitation stations for optimal interpolation. Water Resour. Res. 15: 1841-1850.

(3) Shih, S.F. 1982. Rainfall variation analysis and optimization of gauging systems. Water Resour. Res. 18: 1269-1277.

(4) Hubbard, K.G. 1994. Spatial variability of daily weather variables in the high plains of the USA. Agric. For. Meteorol. 68: 29-41.

(5) Villagra, M.M., O.O.S. Bacchi, R.L. Tuon, and K. Reichardt. 1995. Difficulties of estimating evapotranspiration from the water balance equation. Agric. For. Meteorol. 72: 317-325.

(6) Hopkins, J.S. 1979. The spatial variability of daily temperature and sunshine over uniform terrain. Meteorol. Mag. 106: 278-292.

The Formation and Three-Dimensional Structure of Fingered Flow Patterns in a Water Repellent Sandy Field Soil

Coen J. Ritsema[1], Louis W. Dekker[1], Anton W.J. Heijs[1], and Erik van den Elsen[1]. *[1] DLO Winand Staring Centre for Integrated Land, Soil and Water Research, P.O. Box 125, 6700 AC Wageningen, Netherlands.*

Introduction.   In many different environments one-dimensional models have predicted negligible risk of ground and surface water contamination due to sufficient residence time in the vadose zone. However, it is found now that significant pollution of ground and surface water has occurred where these models predicted none. The cause of these cases of contamination is the occurrence of preferential flow paths in the vadose zone. Such preferential flow paths can be caused by unstable wetting fronts. Unstable wetting fronts begin as a horizontal front which under certain conditions break into fingers as the front moves downward. These fingers facilitate the transport of contaminants to the groundwater at velocities many times those calculated if a stable horizontal front is assumed. The conditions under which fingered flow evolves in field soils are not yet understood, because systematic investigations have been rare. Therefore, it is important to investigate the formation and structure of fingered flow patterns in field soils. Here, evidence is given of finger formation during individual rain events in a water repellent sandy field soil. Additionally, three-dimensional visualizations of fingered flow and water repellency patterns are presented.

Materials and Methods.   The formation of fingered flow patterns has been studied in a 195 cm long and 70 cm deep vertical transect in an extremely water repellent sandy soil near Ouddorp in the coastal dune area in the southwestern part of the Netherlands. The soil, classified as a mesic Typic Psammaquent, was covered by a natural grass vegetation, and consisted of a 9 cm thick humic surface layer on top of a fine dune sand to 200 cm depth (1,2). In situ soil water content measurements were carried out using a stand-alone Trase TDR measurement device to which 98 3-wire probes were connected. The probes were installed within the transect at 4, 12, 20, 30, 40, 55, and 70 cm depth. Per depth, 14 probes were installed horizontally with intervals of 15 cm. Measurement frequency of the probes was once every three hours. Rainfall and groundwater levels were also recorded.

To study the three-dimensional fingered flow and water repellency patterns, ten 120 cm long, 60 wide and 52 cm deep soil blocks were excavated in the same experimental field during a one year cycle. Each block was excavated by taking two hundred 100 cm^3 samples in a fixed grid at seven depths, yielding a total of 1400 samples per block. Sample collection was executed at 0-5, 7-12, 14.5-19.5, 22-27, 29.5-34.5, 37-42, and 47-52 cm depth (3). Each sample was used to determine the soil water content by oven-drying, and to determine the degree of water repellency using the Water Drop Penetration Time Test (4). Three-dimensional visualizations were made using the Iris Explorer (version 2.2) modular visualization software environment of Silicon Graphics Inc. (SGI) (5).

Results and Discussion.   The appearance and disappearance of fingers are shown for three different rain events in the period December 1994 to January 1995 (Fig. 1). The initial soil water content distribution before each rain event is characterized by a relatively wet topsoil and dry dune sand underneath. From the graphs in Fig. 1 it can be seen that two fingers were formed during each rain event, each time at the same locations. Depending on the rainfall amount and duration the sandy subsoil became more or less wetted during the rain showers. Drying of the profile between successive rain events was due to drainage and evaporation. The measurements demonstrate clearly that fingers reoccur at the same locations within this experimental field, and that fingers were formed within a few hours to days indicating their possible accelerating effect on downwardly directed water and solute transport.

For the soil block sampled on September 1, 1992, the isosurface volumetric soil water content of 8.5% is visualized in Fig. 2. From this Figure it can be seen that the vertically directed flow paths started at the layer interface at around 9 cm depth. Absolute soil water content differences between the wet fingers and the surrounding dry soil decreased with depth as water repellency did. Highest degree of water repellency was found at the bottom of the humic topsoil, however spatial differences in water repellency existed in this layer. The starting points of the vertically directed fingers appeared to be related to places with a relatively low degree in water repellency at this depth, indicating that fingers will reoccur at the same sites during new rain events. In arable land, position of the fingers will change in time due to soil tillage practices.

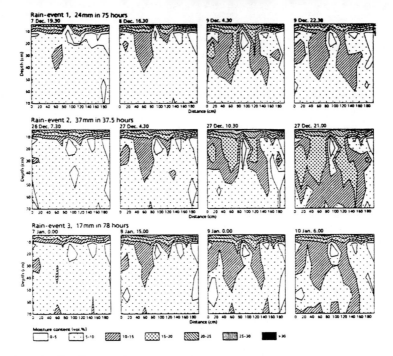

Fig. 1. Formation of vertically directed fingers in the water repellent sandy soil near Ouddorp during three different rain events in the period December 1994 and January 1995.

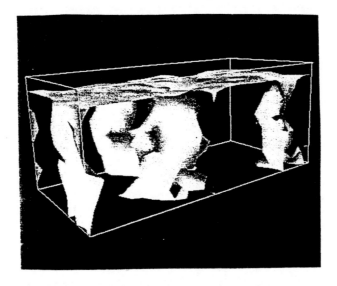

Fig. 2. Three-dimensional visualization of fingered flow patterns in the soil block excavated in the water repellent sandy soil near Ouddorp on September 1, 1992.

Acknowledgments. This work was supported by The Netherlands Integrated Soil Research Programme, and The Environment Research Programme of the European Union. The first organization is thanked for providing the necessary travel funds.

Literature Cited.
(1) Ritsema, C.J. L.W. Dekker, J.M.H. Hendrickx, and W. Hamminga. 1993. Preferential flow mechanism in a water repellent sandy soil. Water Resour. Res. 29:2183-2193.
(2) Ritsema, C.J. and L.W. Dekker. 1994. How water moves in a water repellent sandy soil. II. Dynamics of fingered flow. Water Resour. Res. 30:2519-2531.
(3) Ritsema, C.J. and L.W. Dekker. 1995. Distribution flow: a general process in the top layer of water repellent soils. Water Resour. Res. 31:11987-1200.
(4) Dekker, L.W. and Ritsema, C.J. 1994. How water moves in a water repellent sandy soil. I. Potential and actual water repellency. Water Resour. Res. 30:2507-2517.
(5) Heijs, A.W.J., C.J. Ritsema, and L.W. Dekker. Three-dimensional visualizations of preferential flow patterns at different scales. Geoderma, in press.

Solute Transport in Solid Waste Material

H. Rosqvist[1], C. Öman [2]. [1]*Dept of Water Resources Engineering, Lund University, Lund, Sweden.* [2]*Swedish Environmental Research Institute, Stockholm, Sweden*

Introduction.   The main environmental risk of solid waste landfills is groundwater and/or surface water pollution. The pollutant load depends on the quantity and quality of the water percolating through the landfill. The material composing a landfill originates from a great variety of sources. As a result the landfill becomes highly heterogenous. Research has been carried out to describe solute transport in natural soils, e.g., (1,2). However, solute transport in solid waste is not yet fully understood, e.g. (3). The objective of this study is to determine the variation of hydraulic parameters governing the solute transport in solid waste material. This is done by column experiments and numerical simulation.

Materials and Methods.   A large-scale undisturbed solid waste sample, 1.2 m high and 2.0 m in diameter, was taken in a landfill containing 22-year old crushed household waste (Fig. 1.). A steel cylinder was carefully driven into the solid waste material, while simultaneously excavating the material surrounding the cylinder. Before lifting, a steel sheet was forced under the cylinder and fixed by welding. The cylinder was brought to the laboratory for tracer experiments. Conservative tracers (Lithium bromide and Tritium) were compared to not conservative organic compounds. The organic compounds were assumed to be effected by different processes such as; sorption, microbiological degradation, hydrolysis, and evaporation. The solute was sampled from 28 points at four depths.

The large-scale column experiment was compared to results of two smaller-scale columns of 0.70 m height and 0.53 m diameter filled with crushed, fresh, household waste. The columns were put under a constant water head until steady flow was reached. Then, a tracer pulse containing Lithium bromide and Brilliant blue FCF (4) was applied. After the tracer pulse had infiltrated, the water head was re-established. Solute samples were taken at 4 different depths.

Figure 1. Undisturbed sampling of a large-scale solid waste column.

Results and Discussion.   In Fig. 2 the break through curves (BTCs) for tracer concentrations in the discharge of one of the smaller-scale columns are shown. The correlation between the tracers is high, between 0.96-0.98, and they show a prolonged tailing. The tailing phenomena can be explained by a faster solute transport in larger pore systems and a slower in smaller pores in the matrix. The more extended tailing for the Brilliant blue FCF may be explained by retardation due to adsorption. In Fig. 3, BTCs for Lithium at different depths in one of the smaller-scale columns indicated an increase in tailing and a decrease in peak concentration with depth. No occurrence of preferential flow was observed. Hydrodynamic dispersion coefficients and pore-water velocities for the different tracers as well as at different depths were determined by adapting analytical models to the BTCs.

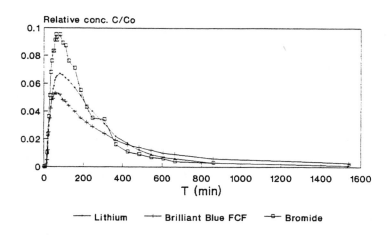

Figure 2. BTCs for the tracers used in one of the smaller-scale columns.

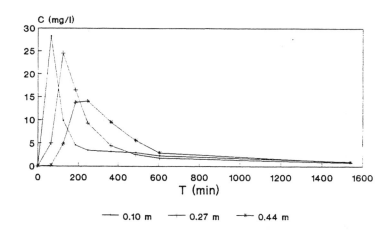

Figure 3. BTCs for Lithium at different depths in one of the smaller-scale columns.

Literature Cited
(1) Nielsen, D.R., van Genuchten, M. Th., Biggar, J. W. (1986) Water flow and solute transport processes in the unsaturated zone. Water Resour. Res., 22(9), 89-108.
(2) Jacobsen, O.H., Leije, F. J., van Genuchten, M. T. (1992) Parameter determination for chloride and tritium transport in undisturbed lysimeters during steady flow. Nord. Hydrol., 23, 89-104.
(3) Vincent, F., Beaudoing, G., Colin, F. (1991) Waste behavior modelling: a numerical model to describe the flow, transport and biodegradation processes. Proc. Sardinia 91, Third International Landfill Symposium, 847-855.
(4) Flury, M., Flühler, H. (1995) Tracer characteristics of brilliant blue FCF. Soil. Sci. Soc. Am. J. 59:22-27.

A Modified Multicomponent Model for Heavy Metals Transport in Soils

H. M. Selim[1], L. Ma[1], Hongxia Zhu[1], I. K. Iskandar[2], and M. C. Amacher.[3] [1]Louisiana State University, Baton Rouge, LA, [2]U. S. Army Cold Regions Research and Engineering Laboratory, Hanover, NH, [3]U.S. Forest Service, USDA, Logan, Utah.

Introduction: The fate of many heavy metal solutes in the environment is of some concern because of their potential reactivity, toxicity and mobility in the soil. Moreover, heavy metals in the soil undergo a series of complex interactions which are governed by soil properties and conditions. Understanding of such complex interactions is a prerequisite in the effort to predict the behavior of heavy metals and mobility in the soil system. To predict the transport of heavy metals in the soil, models that include retention and release reactions of solutes with the soil matrix are needed. Retention and release reactions in soils include ion exchange, adsorption/desorption, precipitation/dissolution, and other mechanisms such as chemical or biological transformations. Adsorption is the process where solutes bind or adhere to soil matrix surfaces to form outer- or inner-sphere solute surface-site complexes.

Objectives: The objectives of this study were to present a modified multicomponent approach for describing the reactivity of heavy metals during transport in soils. The model is based on the assumption that two mechanisms were considered as the dominant retention processes in the soil, namely ion exchange and specific sorption.

Model Assumptions: We considered ion exchange as a non-specific sorption/desorption process. Moreover, ion exchange was considered a fully reversible mechanism and was assumed to be either instantaneous (i.e. equilibrium) or may be considered as a kinetic process. On the other hand, specific sorption was considered as a kinetic process where it was assumed that ions have high affinity for specific sites on matrix surfaces. Furthermore, retention of ions via specific sorption was regarded as an irreversible or weakly reversible process (Selim et al., 1992). The introduction of kinetic processes was based on several observations. In fact, for several heavy metals (e.g. Cu, Hg, Cr, Cd, and Zn), retention/ release reactions in the soil solution have been observed to be strongly time-dependent (Amacher et al. 1988). Several studies indicated that ion exchange is a kinetic process in which equilibrium was not reached instantaneously, e.g., aluminum, ammonium, and potassium. Observed kinetic ion exchange was probably due to diffusion and chemical kinetic processes. It is postulated that in 2:1 type minerals, intra-particle diffusion is a rate-controlling mechanism governing the kinetics of adsorption of cations. Therefore, we extended our model formulation to account for kinetic ion exchange. The proposed approach was analogous to mass transfer or diffusion between the solid and solution phase.

The role of specific sorption and its influence on metal ion behavior in soils has been recognized by several scientists. Sorption/desorption studies showed that specific sorption mechanisms are responsible for metal ion retention for low concentrations. The general view was that metal ions have a high affinity for sorption sites of oxide-mineral surfaces. In the model presented here, we assumed a direct reaction between metal ions in soil solution and specific sorption sites. Others have considered a consecutive type approach for Cd sorption. According to Theis et al. (1988), a set of two second-order reactions were considered; one fully reversible step was followed by an irreversible reaction. Ion exchange as discussed above was not considered. Theis et al. (1988) argued that the amount adsorbed on goethite surfaces was susceptible to migration (via surface diffusion) from primary to secondary surface sites. Other possible mechanisms may include hydrolysis of sorbate at the surface and surface precipitation reactions. In this study the direct reaction and consecutive type approaches were examined.

Materials and Methods: The transport of Cu in Cecil and McLaren soils were investigated using the miscible displacement technique. Plexiglass columns (6.4 cm i.d.x10 cm) were uniformly packed with air-dry soil and were slowly water-saturated. Upon saturation, the fluxes were adjusted to the desired flow rates. A Cu pulse of 100 mg L^{-1} was introduced into each column after it was totally saturated with 0.01 N $MgSO_4$ or $Mg(ClO_4)_2$ as the background solution. Perchlorate as the background solution was used to minimize ion pair formation. The Cu pulse was eluted subsequently with 0.005 M $MgSO_4$ solution. The ionic strength was maintained nearly constant throughout the experiment. In other soil column experiments, similar conditions were used except that no background solution was used in the Cu pulse input solution. This resulted in a condition of variable total ionic concentrations or ionic strength during input pulse application and the subsequent leaching solution. A fraction collector was used to collect column effluent. To obtain independent estimates for the dispersion coefficient (D), separate pulses of a tracer solution were applied to each soil column prior to Cu pulse applications.

Isotherms were measured using standard ion exchange methods where ten-gram samples of soil were equilibrated with Cu and Mg at varying ratios. The samples were shaken for 24 h on a reciprocal shaker with 30 ml of various proportions of $CuSO_4$ and $MgSO_4$ solutions. The solutions were then centrifuged and decanted. For the first two steps (24 h each step) total concentration was 0.5 N followed by four time steps at 0.01 N. Triplicate samples were used for each solution ratio. Adsorbed cations were removed by three extractions with 1 N NaOAc and corrections were made for the entrained solution. Cu and Mg in solution and extractant solution were analyzed by ICP. Based on these Cu-Mg exchange isotherm experiments, selectivity for Cecil and McLaren soils were obtained.

Results and Discussion: When Cu was introduced in Mg background solution with minimum change in ionic strength, Cu breakthrough curves (BTCs) appeared somewhat symmetrical in shape with considerable tailing. Mg BTC showed an initial increase in concentration due to slight increase in ionic strength followed by a continued decrease during leaching. When Cu was introduced in the absence of a background solution, the total concentration considerably decreased. Cu BTC showed a sharp increase in concentration due to chromatographic (or snow-plow) effect (Selim et al. 1992). The Cu peak concentration was 94 mg L^{-1} and the corresponding Mg concentration in the effluent decreased due to depletion of Mg during the introduction of Cu. Mg concentration increased, thereafter, to a steady state level during subsequent leaching, however. This snow-plow effect is a strong indication of competitive ion exchange between Mg and Cu cations which is consistent with our model assumptions.

Literature Cited.
(1) Amacher, M. C., H.M. Selim, and I.K. Iskandar. 1988. Kinetics of chromium (VI) and cadmium retention in soils; A nonlinear multireaction model. Soil Sci. Soc. Am. J. 52,398-408.
(2) Selim, H. M., B. Buchter, C. Hinz and L. Ma. 1992. Modeling the transport and retention of cadmium in soils: Mulltireaction and multicomponent approaches. Soil Sci. Soc. Am. J. 56:1004-1015.
(3) Theis, T. L., R. Iyer, and L. W. Kaul. 1988. Kinetic studies of cadmium and ferricyanide adsorption on goethite. Environ. Sci. Technol. 22:1013-1017.

Method for Observing Interrelationships between Soil Physical Properties and Soil Microbial Colonization

J.S. Selker[1], P. Bottomley[2], J. Garison[1], and T. Sawyer[2]. [1] *Department of Bioresource Engineering and* [2]*Department of Microbiology, Oregon State University, Corvallis, Oregon, 97331.*

Introduction. The sharp divisions between the disciplines of Soil Physics and Soil Biology has lead to a gulf in understanding the profound interactions between the processes which each group considers. Biofouling has received some attention under strongly humanly altered environmental conditions (1). The more subtle processes involved in soil formation, heterogeneity in distribution of water and microbes, and the feedback processes which link these evolving features are poorly understood. To mention the most obvious examples, we note that variations in soil texture effect the distribution and movement of water, and thus biological substrates. As microbes accommodate these factors, they contribute silt sized cells, surfactants, and geologically reactive compounds, effecting the soil physical properties. We seek to explore the following issues: Is the geometry of microbial colonization of vadose materials intrinsically unstable, leading to heterogeneities in both colonization and soil properties? What are the patterns of colonization in the vicinity of textural interfaces in the vadose zone? How might this effect the capillary barrier effects of such features? We present a novel system for non-destructive quantification of joint microbial and soil physical processes based on genetically engineered light emitting bacteria.

Materials and Methods. These experiments are conducted in a thin-slab chamber of packed sand textured quartz media. The chambers employed have two scales. The bench scale chambers are 0.6 m high, 0.5 m wide, and 0.01 m thick. The intermediate scale chamber is 2 m wide, 1.5 m high, and 0.01 m thick. The front and back boundaries of the chambers are glass panels. Fluid distribution is measured non-destructively using the light transmission method (2) with a digital cryogenic CCD imaging camera as the sensor. Distribution of microbial distribution and activity is monitored by colonizing the chamber with Pseudomonas Bacteria modified to include the Lux reported gene cassette in the naphthalene/salicilate degradation pathway (3). The bacteria are robust degraders of naphthalene, and emit detectable light in proportion to their rate of utilization of either of the mentioned substrates. The microbes also grow well on a wide range of alternative substrates, but without activation of the light emission function. The level of light emission is very low (approximately 100,000 photons mm^{-2} sec^{-1} at the chamber surface, requiring 10 minute exposures of equivalent of ASA 200,000 imaging device).

The chamber is filled with any of four grades of pure quartz sand of grades 12-20, 20-30, 30-40 and 40-50 which are known by the trade name Accusand from the Unimin Corp. (Le Seuer, MN). The sand is packed dry flowing into the chamber through a sieved prismatic funnel. Colonization of the chamber is achieved after the sand is packed. The chamber is filled with a solution containing 10^5 cells/ml and drained after 1 hr. The chamber is then twice filled with a nutrient broth, each filling being drained after 24 h. Using this procedure in a 23 deg C room yields a cell density of 10^7 cells/ml of media. Substrate is introduced at the upper boundary of the experiment in aqueous phase.

Results and Discussion.
We are at the stage of having established the protocol for system colonization and light detection. Present results are:
1. The colonization procedure developed provides repeatable uniform initial microbial distribution with adhesion sufficient to allow continuous unsaturated flow experiments over week time scales.
2. Light emission from the cells is sufficiently bright to allow precise characterization of the distribution of the cells, but sufficiently dim to preclude the effective use of standard photographic techniques or direct visual observation.
3. Pseudomonas response from time of substrate application to time of light emission is approximately 30 min, being shorter for colonies previously exposed to naphthalene.
4. At flow rates of 1 chamber pore volume per 20 min, degradation of naphthalene exceeded 90% for influent concentrations from 0.5 ppm to 30 ppm (saturated solution).

The system appears to be very well suited to our application, as well as the broader concerns of understanding soil microbiology. The Lux system has been added to at least 5 organism now, and is in principle easily amended to all bacterial strains using now standard genetic manipulation techniques (4).

Literature Cited.

(1) Vandevivere, P. and P. Bavey. 1992. Saturated Hydraulic Conductivity Reduction Caused by Aerobic Bacteria in Sand Columns. Soil Sci. Soc. Am. J. 56:1-13

(2) Tidwell, V.C. and R. J. Glass. 1994. X Ray and Visible Light Transmission for Laboratory Measurement of Two-Dimensional Saturation Fields in Thin-Slab Systems. Wat. Resour. Res. 30:2873-2882.

(3) King, J.M.H., P.M. Grazia, B. Applegate, R. Burlage, J. Sanseverino, P. Dunbar, F. Larimer, and G.S. Sayler. 1990. Rapid, sensitive, bioluminescent reporter technology for naphthalene exposure and biodegradation. Science 249: 778-781.

(4) USDOE. 1995. Technology and Information Transfer Subsurface Science Program, April, 1995.

Nature and Amount of Spatial Variability of Soil Water at Multiple Scales

M. Seyfried. *USDA-ARS, Boise, ID 83712, USA.*

Introduction. Physically-based hydrological models are largely founded on the results of point or small plot experimental and conceptual work. Many applications of interest are for much larger scales, however. The spatial variability of the natural environment requires alternative approaches to modeling and parameterizing these models. Changing scale may result in a change in the nature of spatial variability, from deterministic to stochastic, for example, which can simplify modeling by reducing the need for distributed data (1). This is the basis for the representative elementary area (REA) concept (2). Scales smaller than the REA can be represented as stochastic "building blocks" for larger scale models. This implies that the nature, and to some extent the amount, of spatial variability is constant with time. In this paper we examine the spatial variability of measured soil water contents to determine if the nature and amount of spatial variability changes with time and scale. Soil water content is of particular interest because it is directly related to processes such as infiltration/runoff and evapotranspiration.

Materials and Methods. All measurements were made within the Reynolds Creek watershed, which has been described previously (3). Measurements were taken for different purposes so that their format is somewhat variable but we believe that they are sufficiently uniform to answer our question. The specifics of sampling are presented in Table 1. In that table, scale refers to the area sampled, the location names refer the subwatersheds in Reynolds Creek and the dates are the number of dates sampled. The $10 \, m^2$ plot was sampled in a grid pattern within the Lower Sheep Creek (LSC) subwatershed. The sampling for the $13 \times 10^4 \, m^2$ LSC subwatershed was on a 30x30 m grid and that for the $26 \times 10^4 \, m^2$ Upper Sheep Creek (USC) subwatershed was done on a 30x60 m grid. Both grids covered the entire subwatershed. The Reynolds Creek (RC) sampling was done at eight locations in roughly 20-sample transects at 20 m sampling intervals.

Results and Discussion. Some critical results (reported as average volumetric percentage) are summarized in Fig. 1. First, the amount of variability, as described by the standard deviation, changes considerably with sampling date and generally increases with the average soil-water content. This is true for all scales. Second, the variability of the $13 \times 10^4 \, m^2$ LSC subwatershed is similar to that observed in the $10 \, m^2$ plot, indicating that most of the variability is at a relatively small scale and that little precision is gained by making extensive measurements. This has important implications for remote sensing of soil-water because it indicates that the typical synthetic aperture radar pixel size is small enough to capture deterministic variability. Third, at low average soil water contents the amount of variability at all scales is similar. It is therefore reasonable to use nonspatial data to characterize large areas at those times. Finally, the amount of variability at larger scales than $13 \times 10^4 \, m^2$ and most average water contents greatly increases with scale. We attribute these increases to known sources of variability such as soil changes that are introduced as the scale increases. Therefore, this variability may be regarded as deterministic and requires more explicit consideration (e.g., a distributed approach). We do not consider it reasonable to generalize the specific results of the trends of variability and soil-water content to different environments. In fact, opposite trends of variability with average water content have been observed elsewhere (4). We do suggest, however, that, in general, increasing scale will result in additional sources of variability and that the REA should be considered in that context.

Conclusion. In general, we found that both the nature and amount of spatial variability of soil-water content vary with time and scale. The increase in variability with scale is attributable to known sources of variability that can be considered to be deterministic. The changes with time are related to the average soil-water content. There are conditions, however, when the variability appears to be stochastic in nature: over all scales at low average soil-water contents, and over all time for scales between $10 m^2$ and $13 \times 10^4 \, m^2$. In addition to being stochastic, the spatial variability it is relatively low under those conditions so that it may not be problematic for model evaluation and parameterization.

Table 1. Data used in spatial variability analysis.

Method	Depth (cm)	Replicates	Dates	Scale	Location
TDR	0-10	12	37	10 m²	LSC
TDR	0-30	83	5	13 x 10⁴	LSC
TDR	0-30	104	5	26 x 10⁴	USC
Gravimetric	0-5	176	4	25 x 10¹⁰	RC

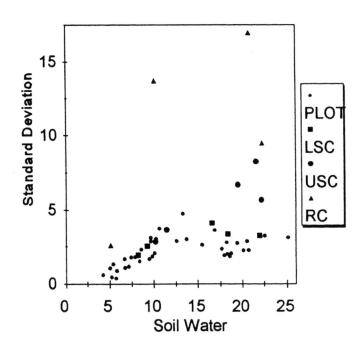

Figure 1. Spatial variability of soil water content for a range of scales.

Literature Cited.
(1) Seyfried, M.S., and B.P. Wilcox,. 1995. Scale and the nature of spatial variability: field examples having implications for hydrologic modeling. Water Resour. Res. 31:173-184.
(2) Wood, E.F., M. Sivapalan, and K. Beven. 1990. Similarity and scale in catchment response. Rev. Geophys. 28:1-18.
(3) Robins, J.S., L.L. Kelly, and W.R. Hamon. 1965. Reynolds Creek in Southwest Idaho: an outdoor hydrologic laboratory. Water Resour. Res. 1:407-413.
(4) Dunin, F.X., and Aston, A.R. 1981. Spatial variability in the water balance of an experimental catchment. Aust. J. Soil Res. 19:113-120.

Non-Invasive Soil Water Content Measurement
Using Electromagnetic Induction

Keith R. Sheets[1] and Jan M.H. Hendrickx[2]. [1]CH2M HILL, PO Box 12681, Oakland CA 94604, USA. [2]Hydrology Program, Department of Earth and Environmental Science and Geophysical Research Center, New Mexico Tech, Socorro, NM 87801, USA.

Introduction. Soil water content data are needed to understand the ecosystems and hydrology of deserts and rangelands. Unfortunately, measurement of soil water content over large areas is a difficult procedure. Common procedures such as "*gravimetry with drying*," "*neutron scattering*," or "*time domain reflectometry*" require a great deal of manpower or are too destructive for repeated measurements at the same location (1). Further, these methods are time-consuming to perform, especially over large areas of extremely heterogeneous rangelands and deserts. The relation between soil water content and soil electrical conductivity has been confirmed by several investigators (2,3,6) which shows that electrical conductivity measurements using electromagnetic induction have the potential for quick non-invasive soil water content measurements. Therefore, the objective of this study is to assess the capability of the electromagnetic ground conductivity meter for monitoring the spatial and temporal variability of soil water content in the Chihuahuan desert.

Materials and Methods. The study was conducted on the New Mexico State University College Ranch, 40 km northeast of Las Cruces, NM. A 2700 m transect has been established here with 90 equally spaced neutron access tubes to monitor soil water content along the transect. All measurements for this study were taken along a 1950 m section of the transect, from station 11 to station 75. The sand and clay content of the soil along the transect is fairly homogeneous. The mean sand and clay content and their standard deviations are: 72.5% ± 4.4 and 13.8% ± 3.9, respectively. We estimate that the average electrical conductivity of the saturation extract in the top 1.5 m of the soil profile along the transect varies between approximately 0.5 and 1.5 dS/m. Climate in the region is characterized by an abundance of sunshine, low relative humidity, an average Class A pan evaporation of 2390 mm per year, and average annual precipitation is 230 mm. Sixty-five measurement stations were established along the transect at 30 m intervals. Soil water content and the bulk soil electrical conductivity of the soil profile (EC_a) were measured simultaneously at each station 16 times between February 1992 and June 1993 (approximately once a month). Soil water content was measured at the 65 stations with a neutron probe at depths of 30, 60, 90, 110, and 130 cm below the soil surface. The water contents at each depth were used to calculate the total amount of water (TWC) in the soil profile at each station to a depth of 1.5 m. The EC_a was measured approximately 10 m south from each soil water measurement station with a Geonics EM-31 Ground Conductivity Meter (5). Measurements were located 10 m away from the neutron probe measurement stations because of the presence of steel supports for rain gauges and thermometers that would affect the readings. Measurements were taken at the soil surface in the horizontal instrument configuration (horizontal/soil surface) and at a height of 89 cm in the vertical instrument configuration (vertical/hip height). It is necessary to standardize field measured EC_a values by conversion to an equivalent electrical conductivity at a reference temperature of 25 degrees Celsius. For the standardization of EC_a we used temperature data at depths of 20, 50, 75, and 100 cm measured near the center of the transect. Linear regression analysis was used to study the relationship between the TWC in the soil profile and the EC_a at each station for each measurement day. We regressed the measured TWC at each station on the EC_a at each station for each individual measurement day (n = 65) to obtain 16 "monthly models."

Results and Discussion. The regression equation coefficients, F and R^2 values are presented by (7). All models are significant at the 0.01 level although the R^2-values remain rather low (11-46%). The residuals of the models are normally distributed with mean zero indicating that the monthly models satisfy the requirements of linear regression analysis. The standard deviations of the residuals for the monthly models are 27 mm TWC for the horizontal/soil surface configuration and 29 mm TWC for the vertical/hip configuration. A mean residual standard deviation of 29 mm over the 1.5 m soil profile represents a water content of 0.019 m^3m^{-3}. This number compares favorable with invasive water content measurement methods when used in the field. To adequately calibrate the monthly models, five neutron probe measurements were needed for the vertical/hip configuration, and nine neutron probe measurements were needed for the horizontal/surface configuration. There was no improvement in the model with the addition of any more stations. Site specific soil water content measurements with the neutron probe (or with another soil water measurement method) will always be required for calibration of the EM method, but these results show that the number of neutron probe soil water content measurements can be greatly reduced. This is important because of the costs involved in installing neutron access tubes, especially in areas with stony soils or caliche layers. Five to nine probes along the transect means one access tube every 200-400 meters. Visualizing each calibration access tube in the center of a square with sides of 400-800 meters, it appears that for our conditions one access tube per 16-64 ha would be needed for calibration. In Fig. 1 we compare the measured and the predicted TWC along the transect during a relatively wet (March 1992) and dry month (May 1993). One striking feature is that the variability of the TWC's predicted on the basis of EM measurements is less than that on the basis of neutron probe measurements. This phenomenon can be explained by the fact that the volume of the EM-31 measurement is at least 4 times larger than the volume of the neutron measurements, thus much of the small scale water content variability is averaged out. An important consequence of this observation is that water content measurements with the EM instrument will often be more reliable than those with the neutron probe. This means that at least

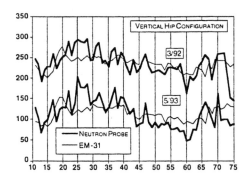

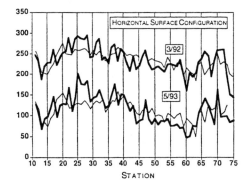

Fig 1. Total water contents for a relatively wet month (top) and dry month (bottom) as measured with the neutron probe and predicted from EM-31 observations with monthly models.

part of the mismatch between predicted and measured values is caused by variability of the neutron probe measurements and not by the failure of the model or EM equipment. Another part of the mismatch can be explained by the fact that the EM-measurements were taken 10 m south of the access tubes, a distance that is certainly large enough to cause discrepancies between neutron probe and EM measurements.

Conclusion.   The results of our study demonstrate that electromagnetic induction is a viable method for measurement of total soil water content (TWC) in the soil profile over long periods of time if the measurements are standardized for the soil temperature. By comparison with the *neutron scattering* method, the *electromagnetic induction* method is quite accurate for the prediction of water content changes over time. The speed and ease of use combined with the accuracy of the measurements make the ground conductivity meter a valuable tool for rapid, non-invasive soil water measurements.

Literature Cited.
(1) Hendrickx, J.M.H., Determination of hydraulic soil properties, in *Process studies in hillslope hydrology*, edited by M.G. Anderson and T.P. Burt, pp 43-92, John Wiley and Sons Ltd., 1990.
(2) Hendrickx, J.M.H., B. Baerends, Z.I. Raza, M. Sadiq, and M. Akram Chaudhry, Soil salinity assessment by electromagnetic induction of irrigated land, *Soil Sci. Soc. Am. J., 56*, 1933-1941, 1992.
(3) Kachanoski, R.G., E.G. Gregorich, I.J. Van Wesenbeeck., Estimating spatial variations of soil water content using non-contacting electromagnetic inductive methods, *Can. J. Soil Sci. 68*, 715-722, 1988.
(4) Kachanoski, R.G., E. De Jong, I.J. Van Wesenbeeck., Field scale patterns of soil water storage from non-contacting measurements of bulk electrical conductivity, *Can. J. Soil Sci. 70*, 537-541, 1990.
(5) McNeill, J.D., Electromagnetic terrain conductivity measurement at low induction numbers. *Tech Note N-6*, Geonics Ltd., Ontario, Canada, pp. 6-8, 1980.
(6) Rhoades, J.D., P.A.C. Raats, and R.J. Prather, Effects of liquid-phase electrical conductivity, water content, & surface conductivity on bulk soil electrical conductivity, *Soil Sci. Soc. Am. J., 40*, 651-655, 1976.
(7) Sheets, K.R., and J.M.H. Hendrickx, Non-invasive soil water content measurement using electromagnetic induction, *Water Resources Research*, in press. 1995.

Numerical Modeling and Field Investigation of Water Flow and the Associated Transport of Oxygen-18 and Chloride in Desert Soils

A. M. Shurbaji. *Department of Land, Air, and Water Resources, University of California, Davis, CA 95616*

Introduction. Vertical profiles of oxygen-18 concentrations have yielded useful information on evaporation and infiltration processes in soils. However, in the field, quantitative interpretation of such profiles has been limited by the restrictions inherent in the quasi-steady-state and transient analytical models available to describe the physical processes (1). A one-dimensional numerical model for simulating transient water flow (vapor and liquid phases), isotope transport (oxygen-18), and heat transfer in the unsaturated zone (ODWISH) (2) is presented. The model can simulate both infiltration and evaporation under fluctuating meteorological conditions and thus should be useful in reproducing changes in field oxygen-18 and chloride profiles. A controlled field experiment was conducted to study second-stage evaporation in soils using oxygen-18 isotope profiles and their relations to moisture and temperature distributions. The data necessarily provide a basis for evaluating the ODWISH model by comparing model predictions with field observations. Field profiles of chloride concentration, and oxygen-18 enrichment in soil-water of an arid New Mexico soil have been studied in an attempt to investigate the rates and mechanisms of recharge and evaporation in desert soils.

Materials and Methods. The controlled field experiment consisted of allowing water to evaporate from homogeneously packed sand columns and periodically sampling to observe changes in the distributions of moisture, temperature, and oxygen-18 enrichment. Water was extracted from the soil for stable-isotope measurement using a vacuum distillation method (3). Moisture content was determined from weight loss of soil during distillation. The oxygen-18 isotopic composition of soil water was measured following the CO_2-H_2O equilibration method (4) using a mass spectrometer. Chloride concentration was determined colorimetrically using mercuric thiocyanite and ferric nitrate solution (5).

Results and Discussion. Figure 1 shows a comparison between model predictions and field data. The match between the measured and calculated distributions of moisture, temperature, and oxygen-18 enrichment is very good. These comparisons indicate that ODWISH simulates the main processes involved in evaporation from soils and the associated isotopic enrichment. Figure 2 depicts the depth distributions of water content, oxygen-18 enrichment, and chloride concentration in a sink hole in New Mexico. At depths between 4.0 m and 4.5 m, the profile has almost uniform water content (=3%) implying gravity drainage and also uniform chloride concentration (C=240.0 mg/l). This condition suggests that quasi-steady-state piston flow can be justifiably assumed and drainage rate can be calculated using the chloride mass balance Eqn. given by: $R\,C = P\,C_o$; where R is the average drainage rate (mm/yr), C is the average chloride concentration of soil-water draining below the root zone (mg/l), P is the average precipitation rate (mm/yr), and Co is the average concentration of chloride in rain water. The drainage rate was estimated as 0.5 mm/yr using C=240 mg/l, P=380 mm/yr, and C_o=0.32 mg/l. The chloride peak corresponds to the maximum depth of extraction of water by roots. The reduction in chloride concentration below the root zone suggest that long-term water infiltration at this location cannot be explained by piston flow through the root zone. Following Sharma and Hughes (5) in their bimodal flow and chloride transport model, the preferential flow via the root zone was determined to represent 85% of the drainage rate below the root zone. The maximum isotopic enrichment (which identifies the evaporation front) occurs at a depth of 27 cm. Assuming that relative humidity is changing linearly from 1.0 at the evaporation front to the average relative humidity and temperature (50%, and 26 °C, respectively) at soil surface, the evaporation rate can be determined using Fick's Law of diffusion (2) as 0.02 mm/day.

Conclusion. Simulations showed that the ODWISH model reproduces reasonably well the experimental distributions of moisture content, isotopic enrichment, and temperature. This comparison, together with the model testing in ref (2), indicate that the hydrologic-isotopic model satisfactorily simulates the processes involved in transient water movement in the shallow unsaturated zone and the associated isotope (oxygen-18) transport and fractionation processes. The isotope data give insight into the evaporation process and water movement in the shallow unsaturated zone while chloride data reveal information about long-term rates and mechanisms of recharge. Analysis using a bimodal flow and chloride transport model provided evidence for preferential flow through the root zone.

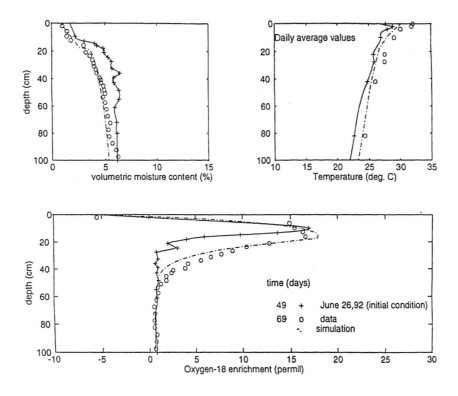

Figure 1. Comparison between model simulations and experimental observations of moisture content, temperature, and oxygen-18 enrichment as a function of depth, after 69 days of evaporation.

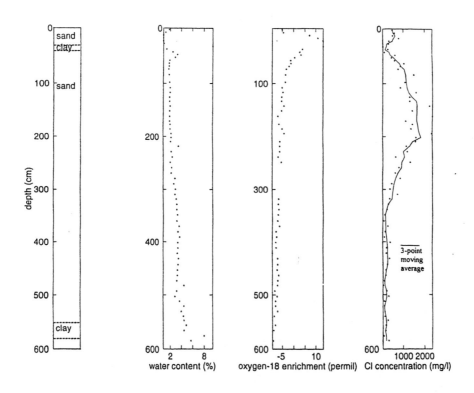

Figure 2. Field depth distribution of water content, oxygen-18 enrichment, and chloride concentration.

Literature Cited.

(1) Barnes, C.J. and G.B. Allison, 1988. Tracing of water movement in the unsaturated zone using stable isotopes of hydrogen and oxygen. *J. Hydrology*, 100:143-176.

(2) Shurbaji, A.M. and F.M. Phillips, 1995. A numerical model for the movement of H_2O, $H_2{}^{18}O$, and ^2HHO in the unsaturated zone. *J. Hydrology* (In press).

(3) Shurbaji, A.M., Phillips, F.M., Campbell, A.R., and Knowlton, R.G., 1995. Application of a numerical model for simulating water flow, isotope transport, and heat transfer in the unsaturated zone. *J. Hydrol.ogy (In press)*.

(4) Roether, W., 1970. Water-CO_2 exchange set-up for the routine ^{18}Oxygen assay of natural waters. *Int. J. Apl. Rad. Isot.* 21: 379-387.

(5) O'Brien, J.E., 1962. Automatic analysis of chloride in sewage. *Waste Engr.*, 33:670-672.

(6) Sharma, M.L. and M.W. Hughes, 1985. Groundwater recharge estimation using chloride, deuterium and oxygen-18 profiles in the deep coastal sands of western australia. *J. Hydrology*, 81: 93-109.

Laboratory Column Experiments Showing the Effects of Solute Concentration-Dependent Surface Tension on Flow in the Vadose Zone

J.E. Smith[1,2] and R.W. Gillham[1]. [1]*Department of Earth Sciences, University of Waterloo, Waterloo, ON, N1G 2L6, Canada.* [2]*currently at Department of Hydrology and Water Resources, University of Arizona, Tucson, AZ 85721, USA.*

Introduction. The isothermal flow equations commonly used in vadose zone hydrology assume surface tension at the air-water interface to be constant. Many organic compounds, including many groundwater contaminants, cause changes in surface tension at the air-water interface as a function of their aqueous concentrations (1). At a given water content, a change in surface tension at the air-water interface causes a proportional change in soil water pressure head (2,3). Therefore, water content, water capacity and unsaturated hydraulic conductivity, which are commonly expressed as functions of pressure head only, become dependant on both pressure head and solute concentration. This causes solute concentration to be a primary variable for flow in the vadose zone. A 1-D numerical model that incorporated changes in surface tension caused by a dissolved contaminant (4), showed that flow in the vadose zone was sensitive to changes in surface tension. The objective of this study was to conduct laboratory column experiments to demonstrate and quantify changes in flow in the vadose zone caused by a dissolved contaminant that changed the surface tension at the air-water interface as a function of solute concentration.

Materials and Methods. Laboratory experiments were conducted in a two meter long, 11.4 cm I.D., vertical sand column packed with Graded Ottawa Sand (ASTM C-109). The column was unsaturated, flowing at steady state, and at the same rate at the start and end of the experiment. The application rate at the top of the column was constant while a water table was maintained at the bottom. Pressure head was measured with Validyne DP-15 pressure transducers attached to porous cup tensiometers installed at seven depths in the column and monitored continually with a microcomputer. Water content was measured using the Time-Domain Reflectometry (TDR) method with rods installed at 13 depths and attached to a common junction through manual rotary switches. The TDR waveform was measured continually using a Tektronic cable tester. The data were transferred to a personal computer and analyzed later using software developed at the University of Waterloo. The column was initially flowing with distilled water before the applied solution was changed to 7% by weight 1-butanol. The surface tension of 7% by weight 1-butanol solution is 35.8% and the kinematic viscosity 133.6% that of distilled water. In addition to monitoring the flow experiments, the tensiometers and TDR rods were used to measure the water content-pressure head relationship. The main drainage curve and main wetting curve were measured with both distilled water and 7% butanol solution. The unsaturated hydraulic conductivity at different water contents was measured with distilled water, 6% butanol and 7% butanol using the steady state flux control method (5).

Results and Discussion. The concentration dependant change in surface tension caused a highly localized flow perturbation to propagate through the column with the solute front. Figure 1 shows pressure head and water content versus time at 38 cm below the top of the column. The initial steady state values are indicated by the dotted lines and the data at late time equal the final steady state values. The changes in pressure head and water content associated with the flow perturbation induced by surface tension changes in the solute front, can be seen clearly. The pressure head increased sharply while the volumetric water content first dropped and then increased as the solute front passed by. This was observed at seven different depths in the upper 150 cm of the column and was more dispersed with depth. The changes in water content were large enough to cause hysteresis in the water content pressure head relationship to have a substantial effect on the final values. One effect was that the pressure head changes were larger than would be predicted by the surface tension change alone. The higher water content at the final steady state, even though the application rate was the same, was due to the higher viscosity of the 1-butanol solution. Another effect observed was depression of the capillary fringe in proportion to the surface tension depression. That caused an associated temporary but substantial increase in effluent rate from the column. A conceptual and/or mathematical model that neglected changes in surface tension as a function of solute concentration would fail to predict the changes in flow observed.

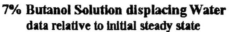

7% Butanol Solution displacing Water
data relative to initial steady state

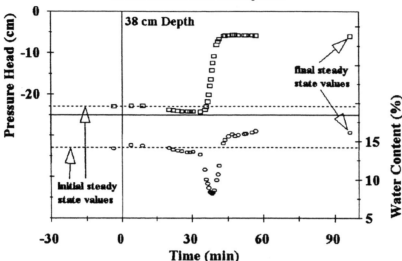

Conclusions. The laboratory results clearly demonstrate that changes in surface tension due to a dissolved contaminant can cause a transient flow perturbation associated with the solute front and also cause depression of the capillary fringe. The effects observed were large and indicate that a significant primary mechanism coupling flow and transport needs to be added to models, both conceptual and mathematical, used to assess organic contaminant problems in the vadose zone.

Literature Cited.
(1) Davies, J.T. and E.K. Rideal. 1963. Interfacial phenomena. 2nd ed. Academic, San Diago, California.
(2) Desai, F.N., A.H. Demond, and K.F. Hayes. 1992. Influence of surfactant sorption on capillary pressure-saturation relationships, in Transport and remediation of subsurface contaminants: colloidal, interfacial, and surfactant phenomena. Chapt 11, edited by D.A. Sabatini and R.C. Knox, American Chemical Society, Washington, D.C.
(3) Milly, P.C. 1982. Moisture and heat transport in hysteretic, inhomogeneous porous media: A matric head based formulation and numerical model. Water Resources Research 18(3):489-498.
(4) Smith, J.E., and R.W. Gillham. 1994. The effect of concentration-dependent surface tension on the flow of water and transport of dissolved organic compounds: A pressure head-based formulation and numerical model. Water Resources Research 30(2):343-354.
(5) Klute, A., and C. Dirkson. 1986. Hydraulic conductivity and diffusivity: laboratory methods. in Methods of soil analysis, part 1, Physical and mineralogical methods. 2nd ed. no 9(1) in series Agronomy. American Society of Agronomy, Madison, Wisconsin, USA.

Effect of Coatings and Fills on Fracture-Matrix Interactions in the Vadose Zone

Wendy E. Soll. *Los Alamos National Laboratory, Los Alamos, NM, 87545, USA.*

Introduction. Disposal of radioactive and chemical contaminants in unsaturated, arid environments is often considered, and has already been implemented in some locations. Many of these locations are in or near fractured porous media. Examples of such sites include Los Alamos National Laboratory (LANL), Idaho National Engineering Laboratory, and Yucca Mountain, Nevada. To predict storage performance or remediation effectiveness at these sites we must be able to incorporate the role of fractures. Unfortunately, incorporating fractures into field-scale models for unsaturated systems is not at all straightforward. An additional complexity is that the fractures are often coated or filled with other materials, either as a result of weathering at the fracture surface, or due to transport down the fracture of particles and fines.

Material and Methods. In this work we have numerically studied the influence of coatings and fills on flow in fractured porous media. The approach we have used is to model a simple system with a single fracture embedded in a homogeneous, isotropic matrix. Simulations were run using FEHM (1), a finite element multiphase flow and transport code developed at LANL. We have so far limited this study to volcanic Tuff matrices - the material with which we are mostly concerned. Two different fractured Tuff systems were used. The Tuff A system was based on local LANL properties (2), while the Tuff B properties were based on Yucca Mountain values, as reported by Nitao (3). For Tuff A we the fracture coating elements were assigned clay properties, reflecting the actual coating materials. For Tuff B we used the matrix properties with reduced conductivity to represent blocking by particulates. For simulations without fracture coatings the coating elements reverted to the matrix parameters to provide a constant fracture width for all simulations. Fracture fills were implemented by assigning clay properties to a section of the fracture.

The entire boundary of the system was no-flow except for the top and bottom of the fracture. All water entered the system through the top of the fracture, and any fluid leaving the system left through the bottom of the fracture. Two inflow conditions were used, a fixed flux rate or a ponded condition. The ponded condition was set up as a saturated top boundary applied for 0.5 days then removed. The constant flux was applied for the duration of the simulations. These simulations set up infiltration events which were beyond even 'extreme' events for the arid to semi-arid environments of concern. Initial saturation conditions depended on the run. All of the Tuff A simulations were run with initial matrix saturation of 0.05. Initial saturations for Tuff B runs varied between 0.15 and 0.65. The fracture saturation was always initially at 0.02.

Simulations were run to look at the sensitivity of flow and fracture-matrix interaction to the existence of coatings and/or fills, and their relative conductivity with respect to that of the matrix. The parameters varied include: presence or absence of a fracture coating, continuity of the coating, and relative magnitude of saturated hydraulic conductivity of the coating with respect to the matrix, initial matrix saturation, presence of fracture fill, location of the upper boundary of the fracture fill, length of the fracture fill, and relative conductivity of the fill with respect to the matrix.

Results and Discussion. The first series of simulations studied the effect of coating conductivity on fracture-matrix flow. With no coating present the matrix rapidly imbibed all water added to the system. With the addition of a coating of 10^{-2} lower conductivity than the matrix, the matrix was still able to imbibe all of the liquid entering the fracture, however the liquid front moved farther than in the no-coating case, and the maximum matrix saturations behind the fluid front were slightly lower. When the coating conductivity was decreased by 10^{-4} below that of the matrix, flow shifted to predominantly in the fracture. Matrix saturation along the entire fracture length increased. These observations are consistent regardless of inflow condition, choice of materials, or initial matrix saturation. For the Tuff B system the influence of initial matrix saturation on coating effectiveness was studied. For low initial matrix saturation, coating conductivity had to be reduced by 10^{-6} before all matrix imbibition ceased, while with higher saturation matrices, 10^{-4} reduction sufficed. The difference is due to the high capillary forces in the matrix at very low saturations.

A second series of simulations were run to study the effects of coating continuity. Results showed that a break in the coating provides a preferred path into the matrix. In the case of fixed injection rate into the fracture, discontinuities in the coating stopped nearly all fracture flow. For a ponded top boundary condition, which effectively increased injection rates by 10^2, water continued to flow down the fracture, as well as being strongly imbibed into the matrix. This was caused by high saturations occurring in the matrix right at the discontinuity. However, once the water source was removed, the water remaining in the system was rapidly imbibed into the matrix.

A third series of simulations was run to analyze the effect of filled fractures. As expected, the fill blocked the liquid from moving down the fracture. For no or high conductivity coatings, the water simply imbibed into the matrix and continued downward as matrix flow. For low conductivity coatings the injected water was effectively blocked from entering the system. For fills that started right at the upper boundary, no water entered the system. Another simulation that was run with a continuous, high saturation, matrix cap on top of the fracture-matrix system also resulted in no movement of water into the fracture, although in this case flow did occur in the matrix.

Figure 1. Water distribution at t=0.5 days for three scenarios with ponded boundary condition and no fracture fill: a) no coating, b) low conductivity coating, c) discontinuous, intermittent coating.

Conclusion. Fracture coatings and fills can significantly influence the movement of water from fracture to matrix in an unsaturated system, but the nature of these changes is not always easily predictable. The simulation results indicate that for these Tuff systems, fractures are only a fast flow path when infiltration is high, such as extreme rainstorms or spring snow melt, and when a fracture coating is present. Fracture fills prevent the movement of water directly down the fracture and force the water to either enter the matrix or pond at the surface. Fracture coatings, unless of extremely low permeability and continuous, are not effective in totally preventing infiltration of water into the matrix, but can act to spread water significantly deeper into the system. Our observations still leave questions unanswered, such as exactly how continuous the coating must be, and for what range of rock types these conclusions are valid. The next step in this study is to try to quantify the rate of matrix imbibition based on these same variables, as well as to look at other characterizations of the fracture-matrix interface and transport processes. These topics will be approached both experimentally, and numerically using microscale and mesoscale modeling.

Literature Cited.
(1) Zyvoloski, G. A. et al., (1995a), Models and Methods Summary for the FEHMN application. Los Alamos National Laboratory document LA-UR-94-3787 Rev. 1.
(2) Rogers, D.B. (1994). Unsaturated Hydraulic Characteristics of the Bandelier Tuff at TA-54, LANL ESH-18 memorandum ESH-18/WQ&H-94-565.
(3) Nitao, J.J. (1991). Theory of Matrix and Fracture Flow Regimes in Unsaturated, Fractured Porous Media, in High Level Radioactive Waste Management, Proc. of the 4th Annual Int'l Conf., Las Vegas, NV:845-852.

Temporal In Situ Changes of Soil Properties
As Affected by Tillage, Position, and Plants

J.L. Starr[1], I.C. Paltineanu[1], D.J. Timlin[2]. [1]Environmental Chemistry Laboratory, BARC, Beltsville, MD 20705. [2]Systems Research Laboratory, BARC, Beltsville, MD 20705.

Introduction. A basic limitation to developing improved cultural practices that will decrease the movement of surface applied chemicals to groundwater is the lack of adequate understanding and quantification of the dynamics of the soil properties that affect leaching. In predictive modeling, many of these soil properties are often incorrectly considered to be constant for a given soil. Some of these changes occur rapidly following plowing, such as soil density and pore-size distribution. Because ground water recharge and chemical leaching commonly occurs from fall to spring seasons (1), these changes need to be known throughout the year. In regions that are implementing no-till cropping, temporal variation studies need to be extended to years to determine if quasi steady state conditions develop. The impact of tillage on nitrate transport to groundwater is not well understood. For example, it is not clear from the literature whether reduced tillage enhances nitrate transport to the groundwater or if more of the infiltrating water flows through macropores by-passing much of the soil matrix, leaving the nitrates (and other agrochemicals) behind. The lack of such basic information along with the increased reporting of unacceptable levels of nitrates in groundwater has led to the designation of "groundwater quality" as the number one research problem by the USDA-ARS (2).

Materials and Methods. In order to study both the short term (within year) and cumulative (year to year) changes in selected physical soil properties, a field site is being incrementally changed from plow-till (PT) to no-till (NT) corn over a four-year period. The experiments were initiated in 1993, plowing all 26 paired-plots (3) in 4.6 by 25 m. In 1994, 1/4 of the plots were in their first year of NT. Three of the *in situ* soil measurements reported here are: a) ponded- and tension-infiltrations, b) shallow bulk densities (D_b) by γ-ray transmission with a Campbell Pacific Nuclear (CPN) strata gage, and c) volumetric soil-water content with a Techtronix 1502B time domain reflectometer (TDR). Modified 12-cm diameter infiltrometers (4) were used for ponded and tension (-3,- 6, -10 cm h) infiltrations. Infiltrations were conducted in the corn rows (between and over corn plants), and between rows (wheel-traffic and nontraffic). The CPN probe has the γ source and detector probes separated by 30 cm, which can be extruded at fixed depth increments of 5 cm to a maximum of 30 cm. Prior to infiltrations, parallel Al tubes were placed on two sides of the infiltration areas to provide true *in situ* comparisons of soil density and infiltrations. When measurements were across plants, the plant stem and brace roots were carefully cut at the soil surface just prior to measurements. Because the stems and brace roots represent positions for high infiltration as they decay, we also measured the diameters and recorded the number of brace intersections along with their spatial positions. Impact of plant canopy on stem vs. through flow of rain were also directly measured from July to Oct. The TDR was attached to a laptop computer via an RS232 serial port for real time measurements. The TDR sensors (3-rod, 25 cm long) were installed vertically in-row and in traffic and nontraffic interows of PT and NT. The field was plowed on 4/27/94, planted on 5/2/94 (replanted 5/31/94), and harvested 10/17/94..

Results and Discussion. Partial results for D_b are presented here for June, Sept., Nov., 1994, and May, 1995 (Fig. 1). In all cases the D_b of NT was greater than PT on the order of 0.2 g cm^{-3}. Besides tillage effects, the importance of field position and time of year for D_b values are evident in Fig. 1. The higher D_b in the wheel traffic interows is maintained to a depth of 25 cm throughout the year for both tillages, though there was some decline in D_b(PT) from June to Sept. In-row and nontraffic interrow D_b values tended to be the same until the May PT (before spring plowing) observations. There seems to have been an overwinter effect to loosen the nontraffic interrow region of PT.

To facilitate comparison of the infiltration data, infiltration rates were normalized to the median in row rates of PT in June. The means of the normalized infiltration rates IR (Table 1) show 30 to 60% greater rates for NT than PT for the non traffic and in-row positions, despite the greater D_b in NT (Fig. 1). In contrast, the tension IRs for these same conditions (data not shown) were higher for PT. Taken together this indicates a greater proportion of large pores in NT with a correspondingly smaller amount of the next smaller pore sizes. The lower IRs for the traffic interrow is consistent with the higher D_b values. The large changes in IR from June to Sept., and again from Sept. to Nov. are especially interesting with implications for runoff and leaching. The drop in infiltrations in Sept. may result from soil roots filling the larger pores (which is not reflected in the "dry" these same roots, plus increased worm activity in the fall season may account for the large rise in Nov., which was previously observed (5). The IR over the year-old corn stalk in the Sept. NT seems to represent a major opportunity for infiltration under heavy rain or irrigation during a time of the year that the rest of the soil surface has a relatively low capacity to take in water. Water that may ordinarily run off can "run down" these natural drains and provide water for the current crop or add to ground water recharge. This may be especially important at sites that are in NT for several years.

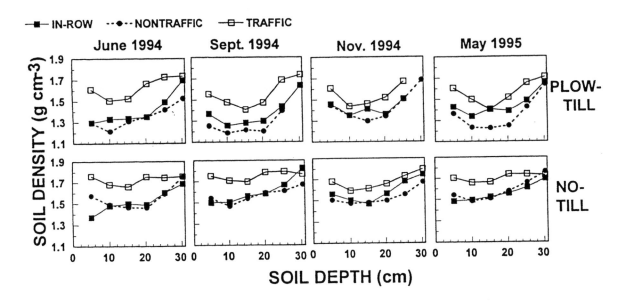

Fig. 1.Temporal *in situ* variation of shallow bulk densities measured by gamma transmission at three plant row positions, in plow- and no-till corn.

Table 1. Ponded infiltration rates at three 1994 dates and positions under plow- and no-till relative to median of PT, June, in-row rates.

POSITION	JUNE		SEPTEMBER		NOVEMBER	
	PT[†]	NT[†]	PT	NT	PT	NT
IN ROW[‡]	1.0	1.6	0.40	0.43	1.1	0.9
Non-traffic	1.2	1.5	0.7	0.2	-	-
Traffic	0.28	0.30	0.28	0.25	0.27	0.12
Yr-old stalk	-	-	-	4.1	-	-

[†] Plow-Till (PT), No-Till (NT)   [‡] No Plants

Literature Cited.

(1) Bergstrom, L. 1987. Nitrate leaching and drainage from annual and perennial crops in tile-drained plots and lysimeters. J. Environ. Qual., 16:11-18.
(2) USDA. 1989. USDA research plan for water quality. Report. 14 pp., ARS-CSRS, Washington, DC.
(3) van Es, H.M., C.L. van Es, and D.K. Cassel. 1989. Application of regionalized variable theory to large-plot field experiments. Soil Sci. Soc. Am. J. 53:1178-1183.
(4) Ankeny, M.D., T.C. Kaspar, and R. Horton. 1988. Design for an automated tension infiltrometer. Soil Sci. Soc. Am. J. 52:893-896.
(5) Starr, J.L. 1990. Spatial and temporal variation of ponded infiltration. Soil Sci. Soc. Am. J. 54:629-636.

Numerical Simulation and Laboratory Tests of LNAPL Displacement Above a Fluctuating Water Table

D.A. Steffy[1], D.A. Barry[1], and C.D. Johnston[2]. *[1]Department of Environmental Engineering, Centre for Water Research, University of Western Australia, Nedlands, Western Australia 6097. [2]CSIRO Division of Water Resources, Private Bag, Wembley, Western Australia 6014.*

Introduction. A series of long-column laboratory tests were conducted to investigate the displacement and entrapment of LNAPL (less dense than water nonaqueous phase liquid) associated with a fluctuating water table. The tests simulated an initially static water table subjected to lateral emplacement of LNAPL and a subsequent rise and fall of the water table. The distribution of the LNAPL throughout the column was measured using dual-energy gamma ray scanning. Results of the laboratory tests were then simulated using a modified form of a finite element immiscible displacement model developed by Katyal et al. (1).

Materials and Methods. A 1-m long, 5.2-cm ID column was packed with a fine-grained calcareous sand. The column was packed in short lifts (2 to 3 cm) in water saturated conditions and subjected to constant stirring to remove air bubbles. The column was then flushed with approximately 10 pore volumes of deaired water under a small (<10 cm) positive head.

In the vadose zone, LNAPL and water pore pressures were measured by a series of hydrophobic and hydrophilic wall tensiometers positioned every 10 cm along the length of the column. The tensiometers were connected through a 12-channel scanivalve to an absolute pressure transducer.

The nondestructive method of gamma ray scanning was used to measure bulk volumetric contents of soil, water and LNAPL. The gamma attenuation method is based on the exponential absorption law of monoenergetic gamma radiation (Ferrand et al., 2). The exponential adsorption law relates gamma ray intensity, I_o^j, and the emerging ray intensity, I^j (the superscript, j, indicates either Cs or Yb), via:

$$I^j = I_o^j \exp(-\mu_w^j \rho_w^j \theta_w x - \mu_n^j \rho_n^j \theta_n x - \mu_s^j \rho_s^j x) \qquad [1]$$

where the subscripts indicate LNAPL (n), water (w) or soil (s), μ is the mass attenuation coefficient, ρ is the density of the corresponding material, θ is the volumetric, and x is the width of the soil column.

For this study, the dual-energy (gamma radiation) sources were Cs^{137} (350 mCi) and Yb^{169} (1,4 Ci). Scanning time was set at 2 minutes with scans taken every 20 mm from 120 to 880 mm above the base and along the length of the column. Because of the short half-life of Yb^{169}, calibration of the mass attenuation coefficient was needed after each scanning sequence. Table 1 lists the mass attenuation coefficients for material pertinent to this study. To improve the attenuation contrast between decane and water, a 7% sodium bromide (NaBr) solution was used and 1-iodoheptane was added to the decane at a 1:9 ratio (Lenhard et al., 3). Because of the gamma scanner could not simultaneously measure both Cs and Yb, scanning was only performed when the column was at a hydrostatic condition.

Results and Discussion. Lateral emplacement of the LNAPL changes the capillary force in the air-water system from 70 dynes/cm surface tension to an air-dedane-water system with interfacial tension of 32 dynes/cm. As a result, the capillary fringe of the water table is correspondingly lowered. Concurrent with the lateral emplacement was the vertical spreading of the decane throughout the length of the column with a resulting residual decane saturation of 21%. Even though the spreading coefficient for decane is small (S = -0.4) in an air-decane-NaBr solution system, there was sufficient attraction between the NaBr solution and decane to enable the vertical spreading of decane throughout the 1-m length of the column.

Numerical simulations were carried out for the conditions of the experiment. These results were compared with the laboratory experiments. There was reasonably good agreement in the changes exhibited by the vadose zone when changing from a two-phase (air-water) system to three-phase (air-LNAPL-water) system.

Table 1. Average and standard deviations of mass attenuation coefficients calculated for 13 calibration runs.

Source	NaBr solution (1/cm)	Decane and 1-iodoheptane (1/cm)	Carbonate sand (1/cm)
Caesium	0.07453 ± 0.00029	0.084357 ± 0.00032	0.070966 ± 0.00275
Ytterbium	0.16761 ± 0.00157	0.399600 ± 0.00899	0.197663 ± 0.00920

Literature Cited.
(1) Ferrand, L.A., P.C.D. Milly, and G.F. Pinder. 1986. Dual-gamma attenuation for the determination of porous medium saturation with respect to three fluids. Water Resour. Res. 22:1657-1663.
(2) Katyal, A.K., J.J. Kaluarachchi, and J.C. Parker. 1991. MOFAT: A two-dimensional finite element program for multiphase flow and multicomponent transport. U.S. Environmental Protection Agency, US EPA 600-2-91/020.
(3) Lenhard, R.J., J.H. Dane, J.C. Parker, and J.J. Kaluarachchi. 1988. Measurement and simulation of one-dimensional transient three-phase flow for monotonic liquid drainage. Water Resour. Res. 24:853-863.

Estimation of Spatial Distribution
Recharge Factors at Yucca Mountain, Nevada

S. A. Stothoff, A. C. Bagtzoglou, H. Castellaw. *Center for Nuclear Waste Regulatory Analyses, Southwest Research Institute, San Antonio, Texas 78238-5166, USA*

Introduction. Determining compliance with the performance objectives for both the repository system and the Geologic Setting (GS) requires prediction of groundwater flow. Since infiltration is the primary source of water in the subsurface, the amounts and locations of infiltration are controlling factors in the movement of groundwater throughout the GS. In fractured, unsaturated rock, such as that found at the Yucca Mountain (YM) site, occurrence of infrequent, high-intensity rainfall events will modify, perhaps drastically, the subsurface flow regime from the one predicted by assuming that all rainfall events have averaged intensities. Indeed, the DOE has concluded that the spatial and temporal distribution of infiltration may be the most important factors influencing groundwater flow path development (5). Deep percolation fluxes are affected by processes active in the near-surface zone, including evaporation, transpiration, liquid water flow, and vapor flow. Each of these processes is governed by several factors. For example, precipitation has been found to vary substantially over the YM region, both spatially and temporally, and winter storms are, in general, more uniform and of longer duration than summer storms (2,3,4). These observations indicate that, especially for summer storms, a spatially uniform precipitation pattern is clearly not applicable, even at the subregional scale. Similarly, evaporation from the ground surface is affected by air temperature, atmospheric vapor pressure, wind speed profile, incident solar radiation, surface soil and rock texture, plant activity, surficial temperature, and surficial moisture content. Many of these factors, such as surficial slope and orientation (related to incident solar radiation), surficial composition, and plant distributions, are or can be mapped at the YM site.

Technical Approach. Site characterization is currently underway in order to investigate the feasibility of siting a geological repository for high-level waste at Yucca Mountain, Nevada. Site characterization involves both field studies and numerical studies, with one aim of the studies being to determine whether probable levels of cumulative radioactive nuclear waste releases meet regulatory standards. Numerical studies have followed a hierarchy, with one extreme considering the entire engineered barrier/geologic setting system in a highly abstracted way and the other extreme considering individual components of the overall system in great detail.

Earlier studies of the Yucca Mountain GS assumed highly idealized material properties and forcing functions; as field studies progress, the models are becoming more sophisticated. To date, most numerical studies of the long-term flow of moisture in the GS have assumed that the infiltration of water past the root zone, herein called recharge, is constant in both space and time, while recognizing that this assumption is questionable. This assumption has the consequence that flux rates at the repository horizon are relatively uniform in space and time. If there are strong focusing mechanisms for moisture flow, there may be considerable impact on repository performance.

A first step towards characterizing the spatial variability of potential recharge was presented by Flint and Flint (1), in which the unsaturated hydraulic conductivity was estimated for the top layer in each grid block in the Wittwer et al. (6) 3-D site model. Using the depth at which seasonal variations of moisture content in the porous media diminish as a reference depth, the present-day moisture contents at the reference depth are used to assess the unsaturated hydraulic conductivity. Applying a unit-gradient assumption to this hydraulic conductivity, and neglecting fracture flow, yields an areally averaged flux of 1.4 mm/yr, with high values of 13.4 mm/yr and low values of 0.02 mm/yr.

In our current work, the idea of using an index to examine the spatial variability of potential recharge is being extended to account for additional characteristics of the near-surface environment at Yucca Mountain. Each characteristic, or potential recharge index, forms a layer in an ARC/INFO Geographic Information System (GIS) database. The ARC/INFO system is classified as a GIS toolset available from Environmental Systems Research Institute (ESRI) Incorporated. GIS systems consist of software applications closely coupled to a geographic database of information. Major GIS functions include spatial operations, data linkage, and geographic databases. Spatial operations enable spatial relationships in data to be explored. GIS systems store data in true relational databases, enabling associations of data sets from a variety of sources to be accessible through a common interface. When all the appropriate factors affecting infiltration are associated using ARC/INFO, potential recharge may be estimated through the implementation of simple phenomenological conceptual models. Using the database manipulation capability of ARC/INFO, sets of layers can be superimposed to find zones with concatenating potential recharge indices, which presumably would be potential high recharge areas. The surface elevation data is at the finest resolution of the data used in the study, at a 30 m × 30 m pixel resolution; all other indices are interpolated to the same resolution.

Potential recharge indices can be classified into categories, depending on whether the index affects precipitation, evaporation, transpiration, infiltration rate, or runoff. Indices based on various properties relevant to fractured porous media have fundamental effects on potential recharge; examples include the matrix hydraulic conductivity, fracture density, fracture aperture distribution, and depth of alluvial cover. Such data are relatively sparse; by assuming that the measured data for each geologic layer is indicative of the layer as a whole, and using available layer outcrop information, estimates of the spatial distribution of surficial media properties are obtained. Simple recharge indices can also be derived purely from topological considerations, such as elevation, ground slope and orientation, and ground concavity. For example, potential insolation is affected by the ground slope, ground orientation, and ridge shadowing, all of which are highly variable in the Yucca Mountain washes. Ground topography also affects the wind profiles, which may have an effect on potential evaporation and transpiration rates.

Evaluating the spatial distribution of the indices feeds back into numerical modeling exercises, by indicating the strength of infiltration focusing mechanisms. Numerical modeling in turn feeds back into the index evaluation process, by assessing the relative importance of each index. Both sets of exercises in turn can be used to assess and guide field site characterization activities at YM.

This report was prepared to document work performed by the Center for Nuclear Waste Regulatory Analyses (CNWRA) for the U.S. Nuclear Regulatory Commission under Contract No. NRC-02-88-005. The activities reported here were performed on behalf of the NRC Office of Nuclear Regulatory Research, Division of Regulatory Applications. The report is an independent product of the CNWRA and does not necessarily reflect the views or regulatory position of the NRC.

Literature Cited.
(1) Flint, A.L., and L.E. Flint. 1994. Spatial distribution of potential near surface moisture flux at Yucca Mountain. *Proc. Fifth Intern. High Level Radioactive Waste Management Conf.* American Nuclear Society: 2,352–2,358.
(2) Hevesi, J.A., J.D. Istok, and A.L. Flint. 1992. Precipitation estimation in mountainous terrain using multivariate geostatistics: Part I: Structural analysis. *J. Appl. Meteor.* 31(7): 661–676.
(3) Hevesi, J.A., A.L. Flint, and J.D. Istok. 1992. Precipitation estimation in mountainous terrain using multivariate geostatistics: Part II: Isohyetal Maps. *J. Appl. Meteor.* 31(7): 677–688.
(4) Hevesi, J.A., D.S. Ambos, and A.L. Flint. 1994. A preliminary characterization of the spatial variability of precipitation at Yucca Mountain, Nevada. *Proc. Fifth Intern. High Level Radioactive Waste Management Conf.* American Nuclear Society: 2,520–2,529.
(5) U.S. Department of Energy. 1992. *Report of Early Site Suitability Evaluation of the Potential Repository Site at Yucca Mountain, Nevada.* Washington, DC.
(6) Wittwer, C.S., G. Chen, and G.S. Bodvarsson. 1993. Studies of the role of fault zones in fluid flow using the site-scale numerical model of Yucca Mountain. *Proc. Fourth Intern. High Level Radioactive Waste Management Conf.* American Nuclear Society: 667–674.

Coherent Structures of the Atmospheric Surface Layer

J. Szilagyi, and M. B. Parlange. *Hydrologic Science, University of California, Davis, CA 95616, USA.*

Introduction. Coherent structures are spatially (and temporally) correlated regions of a turbulent flow which contain a significant amount of the total turbulent energy. They play an important role in the exchange of heat, mass, and momentum at the land–atmosphere interface. Typically, however, coherent structures (or organized events) in the time series are obscured by the superimposed incoherent part of the signal. Wavelet transforms have recently become an instrumental tool for detecting, isolating, and studying these organized events. A new method is proposed to characterize the principal time scale of the coherent structures in the energy containing range of the spectrum. The effect of these structures on the spectral scaling in the inertial subrange is also demonstrated.

Data Collection and Analysis. The three components of the wind velocity as well as the temperature were recorded at 21 Hz by a triaxial ultrasonic anemometer located at 1.5 m above a bare soil surface at the University of California, Davis, Campbell Tract Research Facility under unstable atmospheric stratification conditions. For principal time scale analysis the measurements were block averaged to obtain one value for each second, since the average duration time of the coherent structures is expected to be several tens of seconds. It is possible to estimate the relative importance of coherent structures with different durations using

$$p_m(d=2^{m+1}) = \frac{\sum_{i=1}^{2^{M-m}} (w^{(m)}[i])^2}{\sum_{j=1}^{M} \sum_{k=1}^{2^{M-j}} (w^{(j)}[k])^2} \qquad m=1,\ldots,M \qquad [1]$$

where p_m is the relative importance of structures with duration time d equal to 2^{m+1} (s), $w^{(m)}[i]$ is the coefficient of the wavelet transform at location index i and scale index m, $M=\log_2 N$, and N is the number of measurements. Note that for a white noise process we have

$$\sigma_m^{wn}(d=2^{m+1})> = \frac{\sum_{i=1}^{2^{M-m}} <(w^{(m)}[i])^2>}{\sum_{j=1}^{M} \sum_{k=1}^{2^{M-j}} <(w^{(j)}[k])^2>} = \frac{\sum_{i=1}^{2^{M-m}} 2^{-m}c}{\sum_{j=1}^{M} \sum_{k=1}^{2^{M-j}} 2^{-j}c} = \frac{2^{-2m}}{\sum_{j=1}^{M} 2^{-2j}} = \frac{3}{1} \qquad [2]$$

where <.> denotes the expectation operator. Here we make use of the white noise (wn) property that the squared wavelet coefficients scale with 2^m. In order to obtain a useful estimate of the relative importance of the different coherent structure durations the p_m values for each scale of the white noise process have to be subtracted from the same values of the turbulent measurements. The greater the difference for a given scale m, the more enhanced is the prevalence of the coherent structures for that scale.

Results and Discussion. In Fig. 1. the largest positive value in the differences corresponds to the principal time scale of the structures. In Fig. 1. ten different discrete orthonormal basis functions were applied in order to check whether the choice of the wavelet function has any effect on the results. Fig. 2. displays that during coherent structures (high values of DI, the degree of intermittency) the spatially-local slope of the inertial subrange energy spectrum is greater than -5/3, while in less active regions the slope is found to be less than that. When the slopes of local wavelet spectra are ensemble averaged, an average slope of -5/3 is recovered for the inertial subrange. During the analysis, Taylor's (1938) hypothesis was employed to convert time increments to space increments.

Conclusion. An objective, parameter-free method is proposed for the calculation of principal time scale of coherent structures in the atmospheric surface layer. The identification and study of the coherent structures in turbulent flows is especially important since they locally affect such basic and universal properties as the inertial subrange spectral scaling.

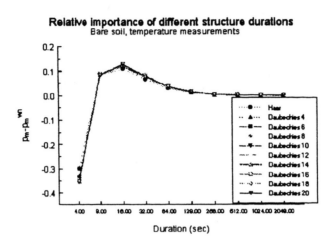

Fig. 1. Relative importance ($p_m - p_m^{wm}$) of different structure durations.

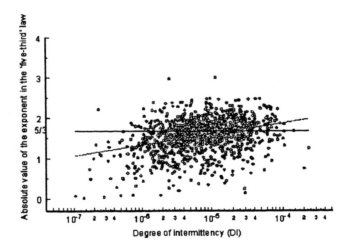

Fig. 2. The exponent of the inertial subrange energy spectrum as a function of coherent structure intensity (DI)

Literature Cited.

(1) Mallat, S. 1989a. A theory of multiresolution signal decomposition: The wavelet representation. Pattern Analysis and Machine Intelligence, 11:674-693.

(2) Mallat, S. 1989b. Multiresolution approximations and wavelet orthonormal bases of $L^2(R)$. Trans. Amer. Math. Soc., 315: 69-87.

(3) Newland, D. E. 1994. Random Vibrations, Spectral and Wavelet Analysis. Longman Scientific and Tec., 477 pp.

(4) Praskovsky, A.A., and S. Oncley. 1994. Measurements of the Kolmogorov constant and intermittency exponent at very high Reynolds numbers. Phys. Fluid, 6: 2886-2888.

(5) Sullivan, P., M. Day, and A. Pollard. 1994. Enhanced VITA technique for turbulent structure identification. Exp. Fluids, 18: 10-16.

(6) Szilagyi, J., G. G. Katul, M. B. Parlange, J. D. Albertson, and A. T. Cahill. 1995. The local effect of intermittency on the inertial subrange energy spectrum of the atmospheric surface layer. Submitted to Boundary-Layer Meteorol.

(7) Szilagyi, J., M. B. Parlange, J. D. Albertson, A. T. Cahill, and G. G. Katul. 1995. An objective method for coherent structure identification at the land-atmosphere interface using orthonormal wavelet expansions. Submitted to Adv. Water Resour.

(8) Taylor, G. I. 1938. The spectrum of turbulence. Proc. Roy. Soc., A, Vol. CLXIV: 476-490.

Integrating 2-Dimensional Ground-Surface Processes in a Saturated-Unsaturated Model to Simulate Different Irrigation Management Practices

K. C. Tarboton and **W. W. Wallender**. *Hydrologic Science, and Biological and Agricultural Engineering, University of California, Davis, CA 95616, USA.*

Introduction. There is a need for a better understanding of processes governing the interaction between irrigation practices and water table elevations, before irrigation management strategies to ameliorate regional problems of elevated saline water tables can be formulated. To better understand interactions between irrigation, evapotranspiration, infiltration, vadose-zone and groundwater flow, field measurements of water table elevation and salinity have been made over a 30 km^2 area near Five Points CA, and a 2-dimensional saturated-unsaturated model (1) has been applied to a vertical transect along the major flow axis, to simulate how irrigation practices on different fields affect the water table elevation below those and adjacent fields. Ground-surface fluxes in the original model (SWMS_2D) were 1-dimensional. In this paper, model development to incorporate 2-D heterogeneous fluxes through the surface boundary is presented. Particular attention is paid to the separation of evapotranspiration (ET) into evaporation and transpiration during the early stages of growth for different crops, and the development of a capability within the model to simulate varying spatial and temporal irrigation practices. The techniques presented, have application at scales larger than field scale, where heterogeneous ground-surface boundary fluxes are an important consideration.

Transect Modeled. A 7 km x 30m vertical transect located on Diener Ranch near Five Points was selected for modeling. Crops grown on this transect include cotton, tomatoes, and seed lettuce in addition to areas left fallow. Both surface and sprinkler irrigation methods have been used at various times on the different crops. Observation wells at the transect ends provide end boundary conditions (BC's) and piezometers at different depths are used to estimate the lower BC. Tile drains are located below several fields with 27 drains intersecting the transect. Model development was carried out on a 1665 m section of the selected transect.

Evapotranspiration partitioning. To model water movement in the root zone, ET needs to be separated into evaporation and transpiration, because evaporation is controlled by the water content and hydraulic gradients at the soil surface, while transpiration can be limited by soil water availability to plant roots. The approach adopted here was to use reference crop evapo-transpiration (ET_o) as an input, from which spatially heterogeneous potential evaporation (E_p) and transpiration (T_p) were determined. ET partitioning depends on the crop pattern and growth stage defined in terms of midday ground shading (G_s) expressed as a function of cumulative ET_o (2) using

$$G_s = G_{sx}(1 + e^{a+bCET_o})^{-1} \qquad [1]$$

where G_{sx} is maximum ground shading, and a and b are crop specific fitting parameters. When $G_s(ET_o)$ data is unavailable, general crop coefficient curves are applied to a particular location to obtain $K_c(ET_o)$ and two points in the nonlinear expression (3) of $G_s(K_c)$ are used determine the crop parameters in [1]. Daily radiation interception (R_i) is determined from ground shading using

$$R_i = 0.0063 + 1.373G_s - 0.392G_s^2 \qquad [2]$$

adapted from (4) and then T_p and E_p are determined using R_i in

$T_p = R_iET_x$	until full canopy cover	[3a]	where	$ET_x = K_{cx}ET_o$	[3b]
$T_p = ET_c$	after full canopy cover	[3c]	and	$E_p = Et_o - T_p$	[3d]

The advantage of using $G_s(ET_o)$ is that different planting dates for a single crop are accommodated with one set of parameters in [1] and at planting $G_s = 0$, hence $E_p = ET_o$ and $T_p = 0$. T_p increases with plant growth, while E_p is reduced accordingly until ET_p exceeds ET_o, after which there is no potential evaporation. After maximum canopy cover is reached, ET partitioning is assumed to remain constant until harvest, while ET_p decreases with the crop coefficient. In the late stages of crop development there is sometimes senescence accompanied by a decrease in radiation interception, however the error introduced by assuming constant ET partitioning in these stages is small as the soil surface is likely dry, hence an increase in E_p changes actual evaporation little. Heterogeneous 2-dimensional crop patterns are accommodated by assigning a specific crop, with associated crop coefficients, ground shading parameters, planting and harvest dates to each ground-surface node to obtain node specific E_p values. T_p is extracted from nodes vertically beneath the surface nodes, according to the crop root distribution at a particular growth stage.

Ground-surface boundary and irrigation. Different irrigation strategies are accommodated by allowing temporally varying, node specific, BC's for each ground-surface node, switching between Dirichlet (specified head) and Neumann (specified flux) BC's. The maximum rate of infiltration or evaporation (E_s) as defined by (5) is limited by the prevailing soil moisture conditions and hydraulic gradient. Each ground-surface node is assigned an irrigation schedule number linked to a file containing information on the date, type, depth (I_d), duration, quality and maximum ponding height (h_p, surface irrigation only) for each irrigation in that schedule. When surface irrigation is specified for certain nodes, the BC on those nodes is set to a Dirichlet BC with h = h_p. Cumulative infiltration (I_c) is tracked at each node until $I_c > I_d - h_p$ after which the specified head is reduced with infiltration until the total irrigation depth has infiltrated. The BC on that node then reverts to a Neumann BC with specified flux (σ). When sprinkler or no irrigation is specified for a ground-surface node, a Neumann BC is used with flux $\sigma = E_p - P$, where the precipitation rate (P) is a node specific sum of irrigation application and rainfall rates. When $\sigma - E_s$ at a node, the BC on that node switches to a Dirichlet BC, with h defined as the product of the excess flux and time step. The Dirichlet BC persists, and h is adjusted at each time step, until $h \leq 0$, whereon the node reverts to a specified flux BC. For Neumann BC's potential evaporation is reduced to an actual value $E_a < E_p$, to maintain $\sigma \leq E_s$ (positive flux out).

Results. In the poster, results are presented showing water table elevations under different heterogeneous crop patterns and irrigation strategies. Ponding on the ground-surface under sprinkler and surface irrigation is illustrated. Sensitivity of water table elevation to changes in crop pattern and irrigation management practice is shown as well as model sensitivity to the location and formulation of the lower BC.

Literature Cited.

(1) Šimunek, J. Vogel, T. and van Genuchten, M. Th., 1994. The SWMS_2D code for simulating water flow and solute transport in two-dimensional variably saturated media. Version 1.2. Res. Report No. 132. U.S. Salinity Lab., USDA, ARS, Riverside, CA, 196p.
(2) Snyder, R.L. 1995. Biometeorologist, Coop. Ext. Univ. California, Davis. Pers. Comm.
(3) Fereres, E., Kitlas, P.A., Pruitt, W.O. and Hagan, R.M., 1980. Development of irrigation management programs. Final Report, Agreement No. B53142, Calif. Dept. of Water Resour., Univ. Calif. Davis.
(4) Arruda, F.B., 1987. Growth of maize and beans as related to plant density, radiation interception and water stress; A simple model. PhD dissert. Dept. Soil Sci. Univ. Calif. Davis, 194p.
(5) Neuman, S.P., Feddes, R.A. and Bresler, E., 1975. Finite element analysis of two-dimensional flow in soils considering water uptake by roots. I. Theory. Soil Sci Soc. Am. Proc. 39, 224-230.

Automated TDR Measurements: Aspects of Software Design

Anton Thomsen. *Danish Inst. of Plant and Soil Science, Research Center Foulum,P.O.Box 23, DK-8830 Tjele, Denmark.*

Introduction. Compared to other aspects of the application of TDR to the measurement of soil water content - e.g. probe design and detailed calibration, software has only received little attention in the litterature. This is unfortunate because well designed software is equally as important for reliable TDR measurements.

Materials and Methods. In order to automate TDR measurements, software for the acquisition and analysis of TDR traces has gradually been developed (Thomsen, 1994). The AUTOTDR software is developed for the Tektronix 1502B/C cable tester interfaced with the rugged Husky FS/2 handheld PC or other IBM compatible personal computer. The AUTOTDR software includes options for the analysis of traces from most of the probe types discussed in the litterature. The included procedures for trace analysis are related to the work of especially Baker and Allmaras (1990) and Heimovaara and Bouten (1990).
The latest version of AUTOTDR includes an option for automated scaling of the parameters defining trace analysis according to the resolution (length) of the acquired trace and the length of the measuring probe. (Scaling for probe length will not be discussed further here). Using default options only the probe length has to be specified by the user. Cable tester settings used for the proportional scaling of default or user specified parameters are retrieved by the PC together with the displayed trace.
In order to test the autoscaling option, moist sandy topsoil was packed into a 25 cm long plexiglass cylinder. Two 20 cm long probe rods were carefully placed 5 cm apart in the soil column. The TDR instrument was connected to the probe rods using an interface cable with a 50/200 ohms balun transformer (Thomsen, 1994). By adjusting the cable tester (distance per division and propagation velocity), the length of the displayed trace was increased in three steps from 1/3 of the screen to full length. For each length resolution, the instrument gain (trace height - not used in parameter scaling) was adjusted in three steps from minimum to full screen height to provide data on the importance of trace height resolution. A total of 25 measurements were made rapidly for each of the 9 resolution combinations.

Results and Discussion. Mean volumetric water content and standard deviation for the 9 series of measurements are shown in Table 1.

Table 1. Mean volumetric water content (%) and standard deviation calculated from 25 measurements within each combination.

Instrument gain (trace height)	Resolution (trace length)					
	High		Medium		Low	
High	18.51	(0.01)	18.41	(0.62)	18.35	(0.03)
Medium	8.51	(0.03)	18.67	(0.06)	18.30	(0.07)
Low	18.71	(0.02)	18.74	(0.07)	18.68	(0.11)

Based on the limited material presented in Table 1, it is concluded that the AUTOTDR software with the autoscaling option produces measurements largely insensitive to trace resolution (length and height) with a low standard deviation.

Conclusions. The apparent success of the parameter autoscaling option is important for both semi-automated and fully automated TDR measurements. Semi-automated measurements using AUTOTDR software are simplified because most variations in probe length and water content can be accommodated without changing instrument settings. Automated measurements are simplified because large fluctuations in water content can be accomodated without data loss if a medium resolution is selected.

Literature Cited.

(1) Baker, J.M. and R.R. Allmaras. 1990. System for automating and multiplexing soil moisture measurements by time domain reflectometry. Soil Sci. Soc. Am. J. 54: 1-6.

(2) Heimovaara, T.J. and W. Bouten. 1990. A computer controlled 36-channel time domain reflectometry system for monitoring soil water content. Water Resources Res. 26: 2311-2316.

(3) Thomsen, A. 1994. Program AUTOTDR for making automated TDR measurements of soil water content. User's guide, vers. 01, January 1994. SP Report No. 38. (Available from the author).

Influence of the Exponent of the Hydraulic Conductiity-
Water Content Relation on Infiltration

J. Touma[1], R. Haverkamp[2] [1]. *Institut Français de Recherche Scientifique pour le Développement en Coopération(ORSTOM), B.P. 5045, 34032, Montpellier Cedex 1, France. [2]Laboratoire d'Etude des Tranferts en Hydrologie et Environnement (LTHE/IMG, UJF, CNRS URA 1512), B.P. 53 X, 38041, Grenoble Cedex, France.*

Introduction. Among the available closed-form expressions to describe soil hydrologic properties, it was shown that those given by (1):

$$\Theta = (\theta - \theta_r) / (\theta_s - \theta_r) = 1 / [1 + (h/h_g)^n]^m , (m = 1 - 2/n) \tag{1}$$

for the water content, θ $(L^3.L^{-3})$, pressure head, h (L); and (2):

$$K = K_s \Theta^n \tag{2}$$

for the hydraulic conductivity, K, $(L.T^{-1})$ water content relationships, satisfy the constraints imposed by the physics of infiltration (3). In these expressions, θ_s $(L^3.L^{-3})$ and K_s $(L.T^1)$ are respectively the saturated water content and hydraulic conductivity, θ_r $(L^3.L^{-3})$, h_g (L), m and η are fitting parameters. The exponent η can be expressed as $\eta = 2/(mn) + 2 + p$, with p either fitted on experimental data or given a fixed value (p = 0, 0.5, 1, 4/3) depending on the adopted model (3 and references herein). Our purpose is to examine the influence of p on infiltration, in order to determine the precision with which this parameter must be determined practically.

Materials and Methods. The Richards' equation:

$$(d\theta / dh).(\partial h / \partial t) = \partial / \partial z [K (\partial h / \partial z - 1) \quad] \tag{3}$$

where t (T) is the time and z (L) is the depth below the soil surface, is solved numerically by an implicit finite difference scheme for two widely used soils: the Yolo light clay (4) and the coarse sand of Grenoble (5), which represent between them a wide range of soil behaviors.The values of the different parameters given in (3) are: θ_s = 0.495 $(cm^3.cm^{-3})$, θ_r = 0.0 $(cm^3.cm^{-3})$, h_g = -19.31 (cm), m = 0.0995, η = 9.143 (p = -1.907) for the clay; and θ_s = 0.312 $(cm^3.cm^{-3})$, θ_r = 0.0 $(cm^3.cm^{-3})$, h_s = -16.39 (cm), m = 0.2838, η = 6.728 (p = 2.2046) for the sand. Results obtained with these values of p are subscripted "r" (reference). The most unfavorable fixed values of p are p = 4/3 for the clay and p = 0 for the coarse sand, to which correspond a relative error of about 30% in η for the two soils. Results corresponding to these values of p (all other parameters unchanged) are subscripted "m" (model).

For both soils, infiltration proceeds under the following conditions:

$$t < 0, z \geq 0; \qquad \theta(z,t) = \theta_i = 0.1\theta_s \tag{4}$$

$$t \geq 0, z = 0; \qquad \theta(z,t) = \theta_s \tag{5}$$

$$t \geq 0, z \to \infty; \qquad \theta(z,t) = \theta_i \tag{6}$$

and lasts for 2 t_{grav}, t_{grav} being the estimated radius of convergence of the series solution of Philip (6) given by:

$$t_{grav} = \{S / [K_s - K(q_i)]\}^2 \tag{7}$$

where S is the sorptivity $(L.T^{-1/2})$, estimated by (7).

Results and Discussion. Figures 1 and 2 present the cumulative infiltration I(t) (a) and the relative difference $(I_m - I) / I$ in percent (b) for the clay and the coarse sand respectively. Even though the relative error in I is important for early times, it drops rapidly to about 15% fot t = 0.1t$_{grav}$, and is less than 10% for t ≥ t$_{grav}$. Computations made with greater initial water content (θ$_i$ = 0.25 θ$_s$) gave smaller relative errors (< 6% and < 5% for the clay and the sand respectively) at the end of infiltration. Thus, any model could be used to estimate the parameter p, and hence the exponent η, since the error in I(t) is less than 10%, which is acceptable for practical purposes.

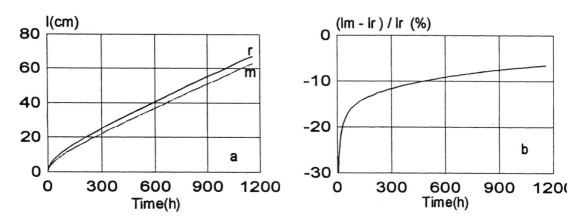

Fig. 1. Cumulative infiltration (a) for the reference (r) and modeled (m) values of η for the clay, and relative difference (b) between the two resulting curves.

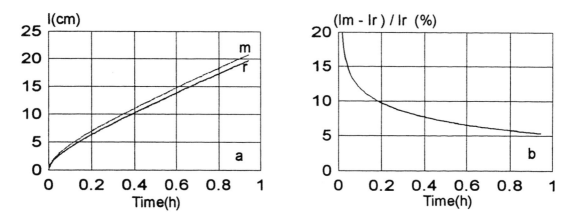

Fig. 2. Cumulative infiltration (a) for the reference (r) and modeled (m) values of η for the coarse sand, and relative difference (b) between the two resulting curves.

Literature cited.
(1) Van Genuchten, M. Th. 1980. A closed-form equation for predicting the hydraulic conductivity of unsaturated soils. Soil Sci. Soc. Am. J. 44 : 892-898.
(2) Brooks, R. H. and A.T. Corey. 1964. Hydraulic properties of porous media. Hydrology Paper no 3, Civil Engineering Dep., Colorado State Univ., Fort Collins, Colo.
(3) Fuentes, C., R. Haverkamp, and J.-Y. Parlange, 1992. Parameter constraints on closed-form soil water relationships. J. Hydrol., 134 : 117-142.
(4) Moore, R.E., 1939. Water conduction from shallow water tables. Hilgardia, 12 : 383-426.
(5) Touma J., G. Vachaud, and J.-Y. Parlange, 1984: Air and water flow in a sealed, ponded vertical soil column : Experiment and model. Soil Sci., 137 : 181-187.
(6) Philip, J. R. 1969. Theory of infiltration. Adv. Hydrosci., 5 : 215-296. Ven Te Chow, Editor.
(7) Parlange, J.-Y., 1975. On solving the flow equation in unsaturated soils by optimization: Horizontal infiltration. Soil Sci. Soc. Am. Proc., 39 : 415-418.

Determining Unsaturated Hydraulic Conductivity of Bandelier Tuff: A Comparison of Characteristic Curve and Ultracentrifuge Methods

H. J. Turin[1], D. B. Rogers [2], and E. L. Vold [3]. *[1] Earth and Environmental Sciences Division, MS F665; [2] Environment, Safety, and Health Division, MS K497; [3] Chemical Science and Technology Division, MS J595; all at Los Alamos National Laboratory, Los Alamos, NM 87545, USA.*

Introduction. Predicting the fate and transport of contaminants in the vadose zone requires knowledge of the unsaturated conductivity (K_{unsat}) of subsurface materials. Conventional laboratory techniques for the direct measurement of K_{unsat} exist (1), but are time-consuming and expensive, especially for measurements at the dry end of the curve. To avoid this problem, a number of prediction methods (2) have been proposed and employed. Typically, these methods are based on more easily measured properties, such as the θ-I characteristic curve. Perhaps the most popular of these methods was proposed by van Genuchten (3), who described the form of the characteristic curve using a few fitting parameters, and then applied Mualem's conductivity model to those parameters. In recent years, a number of researchers have suggested that direct K_{unsat} measurements could be performed more rapidly (and less expensively) by replacing gravity with centrifugal force using an ultracentrifuge (4-5), and a commercial laboratory has recently begun routinely offering these measurements under the trademark UFA.

In support of ongoing environmental restoration and surveillance activities at Los Alamos National Laboratory, numerous K_{unsat} determinations have been performed on Bandelier Tuff, the rock unit underlying Laboratory facilities. Most have been estimated using van Genuchten parameters determined from laboratory-measured characteristic curves, but recently more samples have been submitted for UFA analysis. In order to compare the results of these two fundamentally different methods of determining K_{unsat}, a series of tuff samples was analyzed both by UFA and the characteristic curve method.

Materials and Methods. Thirteen Bandelier Tuff core samples from three different boreholes were submitted for UFA analysis to James Conca (Washington State University - Tri-Cities / Northwest Environmental Services, Testing and Training) and for characteristic curve analysis to Daniel B. Stephens and Associates (DBS&A, Albuquerque). For the three SHB core samples, both labs tested the identical core sample; for the ten G-5 samples, the two labs received adjacent 6-inch core samples. The UFA method measures the moisture content of the core for a given flow rate and centrifugal acceleration. Assuming steady state has been achieved, the hydraulic conductivity at that moisture content can then be calculated. Independent saturated conductivity (K_{sat}) measurements were performed using a falling head permeameter.

At DBS&A, characteristic curve data were collected using a pressure plate for suctions up to about 1000 cm, and a Richards thermocouple psychrometer for higher suctions. van Genuchten parameters were then determined using the RETC nonlinear curve-fitting program (6). Independent K_{sat} measurements were performed by DBS&A using either a constant head or falling head permeameter.

Results and Discussion. Results for three representative core samples are presented in Fig. 1. In each case, the UFA measurements are shown with open circles and the K_{unsat} function predicted by the van Genuchten-Mualem formulation based on the characteristic curve is shown by a solid line. In some cases (e.g. G-5-32), the two methods agree quite well, with the maximum difference of roughly half an order of magnitude occurring near saturation. For other samples, the two methods produce similar shaped curves that diverge at the wet end of the curve. In these cases, the difference between the methods can be attributed solely to differences in the two laboratories' K_{sat} and θ_{sat} determinations. These differences can be compensated for by scaling the van Genuchten-Mualem prediction to the UFA-determined values of K_{sat} and saturated moisture content (θ_{sat}). The resulting scaled predictions are shown in Fig. 1 with dashed lines. In some cases (e.g. SHB-1-168), this scaling process greatly improves the match to the UFA measurements. For a third category of samples (e.g. G-5-60) neither the original van Genuchten-Mualem prediction nor the scaled prediction matches the UFA data well. In these cases, differences in conductivity between the methods can exceed two orders of magnitude at the dry end of the curve, in the vicinity of the 5%-10% *in-situ* moisture contents measured at Los Alamos.

Both of the K_{unsat} determination methods involve major assumptions, and at this point, it is too early to claim that one is correct and the other in error. These results do indicate that more study of these methods is warranted, and that until these differing results are explained, that the new UFA method should not be considered <u>equivalent</u> to established methods. Proposed areas of further investigation include alternative K_{unsat}

K_{unsat} prediction methods (e.g. Brooks-Corey, Burdine formulations), and detailed investigation of dry-end behavior. In particular, the residual moisture content (θ_r) is a parameter in the van Genuchten equations that strongly influences dry-end behavior. This term has come under increasing scrutiny, and may be a weak link in extending K_{unsat} predictions to dry conditions. A recent article has proposed a modification of the van Genuchten equation that replaces θ_r with an adsorption term (7). The present data set provides an excellent test case for this new equation.

Conclusion. Although in some cases, the UFA ultracentrifuge method and an existing characteristic curve-based prediction method yield similar estimates of K_{unsat}, in other cases the two methods diverge greatly. Some of the observed differences are attributable to problems with the measured K_{sat} and θ_{sat} values, but even correcting for this, large variations persist. This study suggests that until these observations can be explained, results provided by the UFA method and characteristic curve prediction methods should not be considered interchangeable. The study also indicates that between laboratories major discrepancies can occur in K_{sat} and θ_{sat} determinations, and that these discrepancies introduce additional error in K_{unsat} results.

Literature Cited.

(1) Klute, A., and C. Dirksen. 1986. Hydraulic Conductivity and Diffusivity: Laboratory Methods. Methods of Soil Analysis, Part 1. Physical and Mineralogical Methods, A. Klute, ed., ASA-SSSA, Madison, WI, 687-734.

(2) Mualem, Y. 1986. Hydraulic Conductivity of Unsaturated Soils: Prediction and Formulas. Methods of Soil Analysis, Part 1. Physical and Mineralogical Methods, A. Klute, ed., ASA-SSSA, Madison, WI, 799-823.

(3) van Genuchten, M. T. 1980. A Closed-form Equation for Predicting the Hydraulic Conductivity of Unsaturated Soils. SSSAJ, 44, 892-898.

(4) Nimmo, J. R., J. Rubin, and D. P. Hammermeister. 1987. Unsaturated Flow in a Centrifugal Field: Measurement of Hydraulic Conductivity and Testing of Darcy's Law. Water Res. Res., 23(1), 124-134.

(5) Conca, J. L., and J. V. Wright. 1992. Flow and Diffusion in Unsaturated Gravel, Soils and Whole Rock. Applied Hydrogeology, 1(1), 5-24.

(6) van Genuchten, M. T., F. J. Leij, and S. R. Yates. 1991. The RETC Code for Quantifying the Hydraulic Functions of Unsaturated Soils. *EPA/600/2-91/065*, U.S. EPA.

(7) Fayer, M. J., and C. S. Simmons. 1995. Modified Soil Water Retention Functions for all Matric Suctions. Water Res. Res., 31(5), 1233-1238.

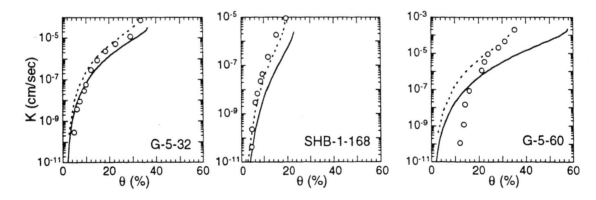

Fig. 1. Unsaturated conductivity curves for three different Bandelier Tuff Samples, determined using three methods. Open circles are ultracentrifuge (UFA) measurements, solid lines are van Genuchten-Mualem predictions based on DBS&A characteristic curve measurements, and dashed lines are the same van Genuchten-Mualem predictions, scaled to UFA-measured θ_{sat} and K_{sat}.

Water and Solute Transport in Structured Clay Soils: Application of Variably Saturated Dual-Porosity Flow Code at Field Scale

G.A.P.H. van den Eertwegh [1] **and J.L. Nieber** [2] : *Wageningen Agricultural University, Dept. of Water Resources, Wageningen, and National Institute of Public Health and Environmental Protection, Bilthoven, The Netherlands.* [2] *University of Minnesota, Dept. of Biosystems and Agricultural Engineering, St. Paul, MN 55108, U.S.A.*

Introduction. Drained clay soils often show cracks. These cracks can be non-permanent due to swelling and shrinkage (1). The presence of cracks, or macropores in general, can cause preferential flow of water and solutes (2). This flow phenomenon can accelerate water transport and leaching of nutrients to drainage systems. The interaction between the soil matrix consisting of micropores, and macropores is very important, as well is the continuity of macropores within the soil (3). To account for water and solute transport influenced at the same time by both the soil matrix and macropores a dual-porosity code has been developed (4). Current work involves applications of the code at the field scale for a cracked heavy clay soil.

Materials and methods. A numerical solution for modeling variably saturated one-dimensional vertical water flow in a dual-porosity porous medium was proposed by (5). The concept of this solution is that the flow domain is separated in two parts. The first part represents the soil matrix with its micropores, the second one treats flow through macropores or fractures. The flow in both systems is assumed to be Darcian. Therefore two Richards' and transport equations are solved simultaneously, coupling the two systems by a transfer term (6). The numerical solution to be used here (4) is able to simulate two-dimensional or three-dimensional axisymmetric flow, taking also water transport by vapor diffusion into account. Soil evaporation and plant transpiration can be modeled. The spatial discretisation has been carried out using the Galerkin finite element method. Finite differences are used to discetize time derivatives using the modified Picard method. The transport solution is derived from the convection dispersion equation (5).

In the Eastern Flevoland polder in the central part of The Netherlands often cracks are present in heavy clay soils. These cracks are caused by ripening of the soil after reclamation of the polder, being a former sea bottom. The soils are drained by subsurface drain tiles and ditches. In the upper soil cracks develop during the growing season which disappear because of tillage operations and swelling in the winter-season. The subsoil cracks are present permanently and are interconnected. Bypass flow and nitrate leaching during the 1989/90 winter period at a site in the polder area has been reported by (7), concluding that bypass flow locally is an important phenomenon and increases the nitrate load to the groundwater and surface water. In 1992 another field experiment has been set up to examine the leaching of pesticides and nutrients (8). The discharge of 3 connected subsurface drain tiles has been measured continuously in addition to collection of water samples to quantify chemical exports.

Results and discussion. The presence of cracks in the soil potentially results in preferential flow and causes the soil to be highly conductive. Measured field drainage in general responds quickly to rainfall excess as shown in Fig. 1. This also occurs in situations in which the subsoil is still partly unsaturated. Within a few days after rainfall events drainage nearly stops. The dual-porosity code has been tested in (4) on experimental flow data from a lysimeter containing a silt loam soil with macropores. The numerical solution is capable of simulating fast response of drainage to application of water at the lysimeter surface. Also rapid chemical outflow is simulated. The results from the lysimeter tests indicate that the observed rapid drainage and solute transport measured in the field experiment will also be represented by the numerical solution. The emphasis of this presentation will be on the comparison of simulated drainage flow and chemical discharge to measured field data. Results of current work on this will be presented at the conference.

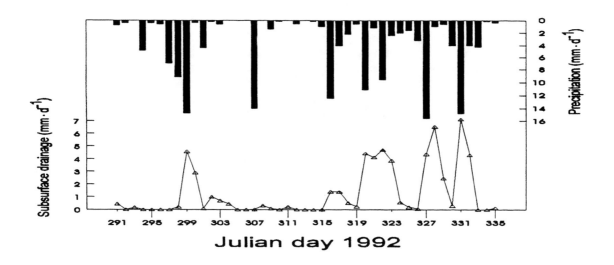

Fig. 1. Precipitation and subsurface drainage Eastern Flevoland field experiment (8).

Literature cited.

(1) Bronswijk, J.J.B. 1991. Magnitude, modeling and significance of swelling and shrinkage processes in clay soils. PhD Thesis Wageningen Agricultural University.

(2) Bouma, J., and L.W. Dekker. 1978. A case study on infiltration into dry clay soils. 1: Morphological observations. Geoderma 20: 27-40.

(3) Beven, K., and P. Germann. 1982. Macropores and water flow in soils. Water Resources Research, Vol. 18, No. 5: 1311-1325.

(4) Nieber, J.L., and D. Misra. 1995. Modeling flow and transport in heterogeneous, dual-porosity drained soils. Irrigation and Drainage Systems 9 (May issue, in press).

(5) Gerke, H.H., and M.Th. van Genuchten. 1993. A dual-porosity model for simulating the preferential movement of water and solutes in structured porous media. Water Resources Research, Vol. 29, No. 2: 305-319.

(6) Gerke, H.H., and M.Th. van Genuchten. 1993. Evaluation of a first-order water transfer term for variably saturated dual-porosity flow models. Water Resources Research, Vol. 29, No. 4: 1225-1238.

(7) Booltink, H.W.G. 1993. Morphometric methods for simulation of water flow. PhD Thesis Wageningen Agricultural University.

(8) Brongers, I., and K.P. Groen. 1994. Emissie van bestrijdingsmiddelen en nutriënten van een akkerbouwbedrijf in Oostelijk Flevoland. Meetresultaten juni 1992 tot juni 1993. Intern rapport Rijkswaterstaat Directie Flevoland (in dutch).

Effects of Soil Type and Water Flux on Solute Transport

J. Vanderborght[1], C. Gonzales[1], M. Vanclooster [1], D. Mallants [1], M. Th. van Genuchten[2], and J. Feyen[1].
[1] *Institute for Land an Water Management, KULeuven, 3000 Leuven, Belgium.* [2] *U.S. Salinity Laboratory, USDA, ARS, Riverside, CA 92501, USA.*

Introduction. Suitable mathematical models and appropriate parameter values are needed to assess the impact of the excessive use of agrochemicals on the subsurface environment. The solute transport process can be conceptualized in differerent ways, leading to different transport models. Based on the evolution of solute dispersion during the transport process, the applicability of two asymptotic transport concepts can be tested (1), i.e., the convection- dispersion equation (CDE), and the convective lognormal transfer function model (CLT) (2).

In order to test the applicability of the CDE and CLT models for different soil types, we monitored the solute breakthrough curves (BTCs) during a quasi steady-state displacement experiment at six observation depths in lysimeters taken from three different soil types. The transport experiments were repeated using different water fluxes so as to asses the effect of the water flux on the model parameters.

Materials and Methods. Three different soil types were selected for the transport experiments: a sandy soil (plaggic Humaquept), a sandy-loam soil (ferrudalfic Udipsamment), and a silt-loam soil (typic Hapludalf). For each soil type, two 100-cm long and 80-cm I.D. undisturbed lysimeters were taken and placed upon a 100-cm long repacked sand column. Six 50-cm long 2-rod TDR probes were inserted horizontally at six different depths (7.5 cm, 22.5 cm, 37.5 cm, 50 cm, 70 cm and 90 cm) in each lysimeter.

For the first displacement experiment, a quasi steady-state water flow regime was established by applying for 1 month a daily dose of 1 cm solute-free water to the lysimeters. Next, a $CaCl_2$ solution was applied continuously until the initial soil solution was completely replaced by the tracer solution. As soon as the salt was removed by leaching with solute-free water, a second transport experiment was carried out by applying a daily dose of 0.5 cm of tracer solution. Time series of moisture content and resident concentrations, C_r (x,t), were obtained at different depths in the soil profile using TDR (3) and appropriate calibrations.

The CDE model parameters D and v were obtained by fitting the solution of the CDE equation for a third type input boundary condition to the observed relative resident concentrations, $c_r(x,t)$ (4). The CLT model parameters, σ and μ, were obtained by fitting the following expression to the observed $c_r(x,t)$ data following a step input of solute (5):

$$c_r(x,t) = \frac{1}{2} \left(1 + erf\left(\frac{\ln(t) - \mu - \sigma^2}{\sigma\sqrt{2}} \right) \right)$$

[1]

Results and Discussion. Based on the CDE model parameters, D and v, we calculated the dispersivity, $\lambda = D/v$, at six depths for all soils. As can be seen from Fig. 1a, λ increased with depth in both the sandy and silt-loam soils, whereas λ remained constant with depth in the sandy-loam soil. The σ parameter remained fairly constant with depth in the sandy and silt-loam soils, whereas σ decreased in the sandy-loam soil (Fig 1b). This result indicates that the solute transport process at the lysimeter scale in the sandy and silt-loam soils are better conceptualized by a stream tube model (CLT) in which the solutes are assumed to move at different velocities in isolated stream tubes. The solutes in the sandy-loam soil on the other hand, are well mixed so that the transport in this soil can be better described using the CDE.

The parameter σ reflects the variance of the solute travel time and increases with the water flux (Fig. 2) thus indicating that the flow regime changes when the water flux changes. Under unsaturated conditions, an increase in water flux will probably lead to the addition of new stream tubes which previously were not conducting water and which had remained dry at the lower flux. As a result, variations in the advection velocity or σ increased with increasing water flux.

Conclusions. Solute transport at the lysimeter scale in soils exhibiting heterogeneity due to local variations in organic carbon content and texture (Humaquept), or due to macropores (Hapludalf), was better described using a stochastic-convective transport concept (CLT). By comparison, the convective-dispersive transport concept (CDE) appeared more appropriate for the more homogeneous, structureless soils (Udipsamment). Because the top boundary condition or the waterflux can significantly influence the values of the solute transport parameters, the water flux used for determination of the parameters should resemble the flux prevailing under natural conditions.

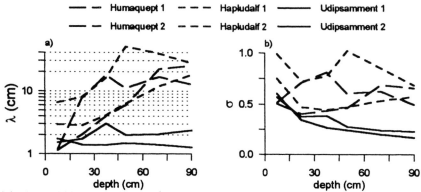

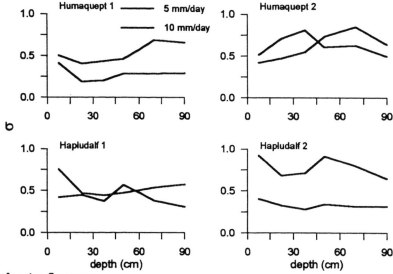

Fig. 1. a) Dispersivity λ, and b) σ as a function of depth for different soil types and for a water flux of 1 cm/day

Fig. 2. The effect of water flux on σ

Literature Cited.
(1) Jury, W.A. and K. Roth. 1990. Transfer functions and solute transport through soil: theory and applications. Birkhäuser Verlag Basel.
(2) Jury, W.A. 1982. Simulation of solute transport using a transfer function model. Water Resour. Res. 18:363-368.
(3) Mallants D., M. Vanclooster, M. Meddahi and J. Feyen. 1994. Estimating solute transport in undisturbed soil columns using time-domain reflectometry. J. Contam. Hydrol., 17:91-109.
(4) Parker, J.C. and M. Th. van Genuchten. 1984. Determining transport parameters from laboratory and field tracer experiments. Bulletin 84-3. Virginia Agric. Exp. Sta., Blacksburg.
(5) Vanderborght J., M. Vanclooster, D. Mallants, J. Diels and J. Feyen. 1995. Determining convective lognormal solute transport parameters from resident concentration data. Soil Sci. Soc. Am. J. Submitted

Modelling Transient Water Flow and Solute Transport Using Alternative Methods to Obtain Soil Hydraulic Properties

K. Verburg[1], W.J. Bond[2], M.E. Probert[3], H.P. Cresswell[2], B.A. Keating[3], K.L. Bristow[4]. [1] *CSIRO Division of Soils, Brisbane,* [2] *CSIRO Division of Soils, Canberra,* [3] *CSIRO Division of Tropical Crops and Pastures, Brisbane,* [4] *CSIRO Division of Soils, Townsville, Australia.*

Introduction. Measurement of hydraulic properties for soil-water-solute balance modelling can be a daunting task, and there is a need for simple methods to estimate these properties. A large number of studies have used properties such as texture and bulk density to predict hydraulic parameters, but the evaluation of such predictions is usually limited to a statistical comparison with measured parameters. In this study we go one step further and assess the effects of alternative methods for estimating the hydraulic properties in terms of the prediction of water and solute transport. We utilise two models, SWIM and APSIM SoilWat, to simulate a field bromide leaching experiment using different levels of hydraulic parameterisation, and compare the predictions with field observations.

Description of dataset. Solution of KBr were applied uniformly to twelve $1 m^2$ plots which were then subjected to natural rainfall. The plots were sampled 1, 2, and 6 months later. Hydraulic properties were measured in situ using the instantaneous profile method, except the 15 bar water content which was measured in the lab. The soil was an alfisol, with distinct textural layering.

Models and input data. SWIMv2 (Soil Water Infiltration and Movement) simulates infiltration, evapotranspiration, redistribution, and unsaturated solute transport based on a numerical solution of the Richards' equation and the advection-dispersion equation (10). In the simulations, dispersion was set equal to diffusion. SoilWat is a multi-layer, cascading water balance module that is part of the Agricultural Production Systems sIMulator (APSIM) (7). It has its origins in the water balances of CERES-Maize (5) and PERFECT (6). The soil water processes are described in terms of lower limit, drained upper limit, and saturation. SoilWat uses a "mixing" algorithm for solute leaching. Table 1 describes the methods that were used to obtain the hydraulic properties for the two models. In the SWIM simulations the smoothed Campbell water retention and hydraulic conductivity curves were used (1,4). The soil hydraulic parameters obtained with the different methods varied considerably (e.g., for 10-50cm depth: $\theta s=0.27-0.35$, b=3.2-6.2, $\psi e=2.3-13.5$ cm, Ks=3.6-8.0 cm/h, LL=0.056-0.13, DUL=0.15-0.26, SAT=0.26-0.29).

Table 1. Methods used to obtain soil hydraulic properties

Case	Method
SWIM0	measured water retention and hydraulic conductivity curves
SWIM1	water retention curve estimated from texture and bulk density (11), θs from 0.93*porosity, Ks from texture (8), K(θ) using (1)
SWIM2	Campbell b predicted from slope of cumulative particle size distribution (2), θs from 0.93*porosity, ψe from one measured $\psi(\theta)$ point, Ks from texture (8), K(θ) using (1)
SWIM3	water retention curve fitted through two measured points (15 bar and 50 cm suction) (3), θs from 0.93*porosity, Ks from texture (8), K(θ) using (1)
SWIM4	water retention curve and Ks estimated from inverse method applied to draining profiles (9)
SoilWat0	LL (15 bar), DUL (100 cm) and SAT (saturation) from measured hydraulic properties
SoilWat1	LL, DUL, and SAT estimated from texture and bulk density (5)
SoilWat2	LL, DUL, and SAT calibrated using draining profiles

Results and Discussion. The average bromide distribution at the second sampling time is shown in Fig. 1 as 95% confidence intervals about the mean concentration. The agreement between the five SWIM predictions and the measurements is very good. All five cases gave adequate predictions of both the peak position and the spreading of the pulse. This suggests that the average soil water flux and storage was predicted with sufficient accuracy, despite the fact that the hydraulic input parameters differed considerably. Similar results were found for the predictions at the other sampling times and for predictions of water content profiles.

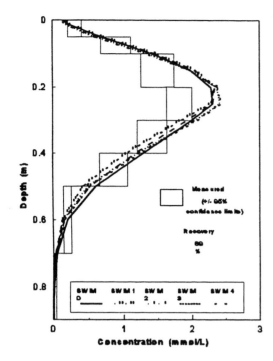

Fig. 1. Measured and predicted solute
profiles at 2 month sampling.

SoilWat predicted the leading edge of the bromide pulse, but did not leach bromide from the surface layers in line with the observed data. The three SoilWat predictions were, however, very similar.

Analysis of the simulations showed that the soil profile was quite wet during the bromide experiment, with matric potentials ranging between -20cm and -90cm for all depths except for the top 10 cm. The results suggest that either the hydraulic property functions were similar over this limited range of matric potential (even though the hydraulic input parameters varied considerably), or else, that the models are relatively insensitive to their hydraulic parameterisation for the conditions of these simulations. In evaluating alternative methods to obtain hydraulic properties it is important to consider the ranges of matric potentials occurring in the experiment. The case presented here is a preliminary study, considering only a single scenario. Further research is currently underway to determine a) the sensitivity of solute transport predictions to hydraulic input parameters, and b) the general applicability of these findings to a range of scenarios. If generally applicable, then this will increase the attractiveness of simulation modelling.

Literature Cited.
(1) Campbell, G.S. 1974. A simple method for determining unsaturated conductivity from moisture retention data. Soil Sci. 117:311-314.
(2) Chang, H. and G. Uehara. 1992. Application of fractal geometry to estimate soil hydraulic properties from the particle-size distribution. p 125-138. In M. Th. van Genuchten, F.J. Leij, and L.J. Lund (eds.) Proc. Int. Workshop on Indirect methods for estimating the hydraulic properties of unsaturated soils. University of California, Riverside, CA.
(3) Cresswell, H.P. and Z. Paydar. 1995. Water retention in Australian soil. I. Description and prediction using parameteric functions. (in preparation)
(4) Hutson, J.L., and A. Cass. 1987. A retentivity function for use in soil-water simulation models. J. Soil Sci. 38:105-113.
(5) Jones, C.A., and J.R. Kiniry. 1986. CERES-Maize: A simulation model of maize growth and development. Texas A&M University Press, College Station, Texas.
(6) Littleboy, M., D.M. Silburn, D.M. Freebairn, D.R. Woodruff, G.L. Hammer, and J.K. Leslie. 1992. Impact of soil erosion on production in cropping systems. I. Development and validation of a simulation model. Aust. J. Soil Res. 30:757-774.
(7) McCown, R.L., G.L. Hammer, J.N.G. Hargraves, D.L. Holzworth, and D.M. Freebairn. 1995. APSIM: A novel software system for model development, model testing, and simulation in agricultural systems research. Agricultural Systems (in press).
(8) Rawls, W.J. and D.L. Brakensiek. 1989. Estimation of soil water retention and hydraulic properties. p. 275-300. In H.J. Morel-Seytoux (ed.) Unsaturated flow in hydrologic modelling theory and practice. Kluwer Academic Publishers.
(9) Ross, P.J. 1993. A method of deriving soil hydraulic properties from field water contents for application in water balance studies. J. Hydrol. 144:143-153.
(10) Ross, P.J., K.L. Bristow, S.W. Bailey, and K.R.J. Smettem. 1992. Using SWIMv2 to model soil water and solute movement, p.83, In Agronomy Abstracts. Proc. 1992 Annual Meetings of the American Society of Agronomy, Minneapolis, U.S.A..
(11) Williams, J., P.J. Ross, and K.L. Bristow. 1992. Prediction of the Campbell water retention function from texture, structure, and organic matter, p. 427-441. In: see reference (2).

Effects of Air Entrapment on Water Flow

Z. Wang and J. Feyen. *Institute for Land and Water Management (ILWM), KULeuven, Vital Decosterstraat 102, 3000 Leuven, Belgium. (E-mail: Zhi.Wang@agr.kuleuven.ac.be)*

Introduction. During water infiltration into the vadose zone over a broad surface area, soil air is displaced and compressed ahead of the wetting front. The effect of air entrapment on water flow has been ignored in the modeling of water flow. Nevertheless, air compression ahead of the wetting front had been frequently noted to affect infiltration rate, soil moisture characteristic, hydraulic conductivity, water table fluctuation and especially the appearance of preferential flow (1,2,3,4,5). Despite all previous investigations, there has been a lack of specially designed quantitative experimentation and theoretical analyses of the basic relations between air and water in the vadose zone. The objectives of this work are: to investigate the effect of air entrapment on infiltration rate; to determine the basic relations between water inflow and air outflow; and to present a theoretical analysis for wetting front instability and the appearance of preferential flow with air compression ahead of the wetting front.

Materials and Methods. To quantitatively detect and measure the simultaneous flow of water and air during unsaturated infiltration, we purposefully designed a laboratory experimental setup which consists of a transparent simulation column (8.6 cm I.D.), a modified tension infiltrometer, and an air flowmeter (for collecting and measuring the rate of air outflow). A dry homogeneous loamy sand was used and packed to a bulk density of 1.60 ± 0.02 g/cm^3. Infiltration tests under constant water heads of -10, -5, 0, 3, 5, 6, 8, 10 cm and variable heads of -10~0, 0~5 and 0~10 cm were conducted. The soil air pressure was measured using water manometers (3 mm I.D.).

Results and Discussion. Air pressure in an air-confined column generally increases with time while fluctuating locally. At every maximum pressure of the fluctuation the entrapped air breaks into the open air, creating vertical and horizontal air channels in the wetted layer, whereas at every minimum pressure the air venting is closed by re-saturation of the top layer. This alternating pattern of air breaking and air closing has been a common phenomenon in all the air confined experiments. The maximum air pressure is introduced here as "air-breaking value", H_b, and the minimum pressure as "air-closing value", H_c. H_b and H_c can be determined as:

$$H_b = Z_{min} + h_0 + h_{ab} \qquad [1]$$

$$H_c = Z_{min} + h_0 + h_{wb} \qquad [2]$$

where, Z_{min} is the minimum wetting depth; h_0 is the surface water head; and h_{ab} and h_{wb} are respectively the "air bubbling value" and the "water-bubbling value" of the material, determined as the suction heads corresponding to the inflection points on the drying-retention and the wetting-retention curves, respectively. h_{wb} corresponds to the "capillary fringe" of the material. It is expected therefore that when h_a exceeds H_b, the entrapped air will break through the saturated layer, leading to a decrease in h_a and an increase in water inflow rate i_w, air outflow rate i_a and the wetting depth Z_{min}. Until h_a falls below H_c, the escape of air will be stopped and h_a will increase again towards generally a higher H_b value. This cyclic process repeats until the end of infiltration.

In air-confined columns, it appeared that: (a) the water inflow rate i_w initially increases with the increasing h_0, then begins to decrease as h_a increases; (b) h_a generally increases with the increasing wetting depth; (c) i_a and i_w are inversely proportional to h_a which is mainly dependent on the stability of h_a; and (d) a minor decrease in h_0 may introduce an immediate drop in h_a and an abrupt increase in i_a and i_w. In the air-escaping columns, however, h_a was almost zero. i_w decreased initially and then kept at a much higher base flow rate, 3-10 times that of air-confined conditions in the loamy sand. In all the experiments the air outflow rate i_a, after a short period of infiltration, is approximately equal to the rate of water inflow i_w, demonstrating immiscible displacement between air and water in the vadose zone.

$$F(h) = h_{aL} - h_{cL} - h_0 \ > 0 \tag{3}$$

where h_{aL} is the air pressure ahead of the wetting front; h_{cL} is the suction head at the wetting front; h_{aL} is a function of both the depth of wetting front, L, and the depth of the air barrier (e.g. the groundwater table), B. During isothermal infiltration into a homogeneous medium the soil air is initially at the prevailing barometric pressure, h_b ($\approx$10 m of water), a critical unstable depth, L^*, is obtained as:

$$L^* = B \ [1 - h_b \ / \ (h_b + h_{cL} + h_0)] \tag{4}$$

A graphical expression of [4] showed that with air compression ahead of the wetting front almost all infiltration flows into unsaturated porous media are subject to wetting front instability and flow fingering. It was proved from the air confined column experiments that shortly after the wetting front was predicted by [3] to be unstable, a "finger" emerged from the initially flat wetting front and ran down quickly, bypassing large matrix of the dry sand material. In the air-escaping columns, however, no fingers were detected.

It can be concluded from this work that air entrapment during unsaturated infiltration strongly affects the rate and route of water flow. Infiltration rate is reduced by several folds giving rise to greater flood risk from a heavy rain. The groundwater may be contaminated earlier than expected because of preferential flow.

Literature Cited.

(1) Fayer, M.J. and D. Hillel. 1986. Air Encapsulation: II. Profile water storage and shallow water table fluctuations. Soil Sci. Soc. Am. J. 50: 572-577.
(2) Linden D.R. and R.M. Dixon. 1973. Infiltration and water table effects of soil air pressure under border irrigation. Soil Sci. Soc. Am. J. 37: 94-98.
(3) Peck A.J. 1965. Moisture profile development and air compression during water uptake by bounded porous bodies: 3. Vertical columns. Soil Sci. 100: 44-51.
(4) White I., P.M. Columbera and J.R. Philip. 1977. Experimental Studies of wetting front instability induced by gradual changes of pressure gradient and by heterogeneous porous media. Soil Sci. Soc. Am. Proc. 41: 483-489.
(5) Wilson L.G., and J.N. Luthin. 1963. Effects of air flow ahead of the wetting front on infiltration. Soil Sci. 91: 137-143.

Monitoring Ionic Solutes under Transient Soil Conditions Using Time Domain Reflectometry

Jon M. Wraith, Patricia D. Risler, and Hesham M. Gaber. *Plant, Soil & Environmental Sciences Department, Montana State University, Bozeman, MT 59717.*

Introduction. A number of investigators have used time domain reflectometry (TDR) to monitor transport of ionic solutes during steady flow through soils. This ability is based on measurement of the bulk apparent soil electrical conductivity (σ_a), which changes in response to movement of ionic solutes past the TDR probe location (1,2,3). In order to apply this monitoring technique in a useful manner to typical field conditions, the substantial influences of changing soil water status and temperature on σ_a must be simultaneously compensated for. This will then allow direct estimation of the soil solution electrical conductivity (σ_w), which is a measure of total ionic concentration. Potential applications include field tracer experiments, monitoring root zone soil water and fertilizer status, and monitoring leakage from chemical containment facilities. Of particular advantage is the ability to non-invasively measure at a large number of soil locations with very high resolution in space and time. This approach should contribute to resolution of issues related to miscible displacement and the spatial/temporal structure of chemical transport processes in the field, topics of particular relevance to the work of Drs. Biggar and Nielsen.

Materials and Methods. Miscible displacement experiments were conducted in the laboratory (4) and field (5) using KBr as a conservative tracer. Transient water flow conditions were imposed in repacked laboratory soil columns by programming a datalogger to control a power relay to a syringe pump. In the field we intermittently ponded the soil surface to achieve transient flow. We used a linear relationship (6) to relate bulk soil electrical conductivity to that of the liquid phase:

$$\sigma_a = T\theta\sigma_w + \sigma_s \qquad [1]$$

where T is a transmission coefficient to account for the tortuosity of the electrical current flow paths, θ is volume water content, and σ_s is surface conductivity of the soil particles. Soil-specific values of T and σ_s were calibrated by alternately wetting and drying the soil at constant known σ_w, while simultaneously measuring θ and σ_a using TDR. Calibration was thus performed in the same soil volumes as used in subsequent transport experiments. Soil temperature was held constant during the laboratory trials, and was measured in the field adjacent to TDR probes for compensating measured σ_a. Pulse concentrations were selected so that soil σ_w levels were within the range encountered in non-saline agricultural soils. TDR-estimated breakthrough curves (BTCs) were compared to those based on soil solution Br concentration (ion-specific electrode) and σ_w (EC meter). We collected effluent fractions in the laboratory, and used vacuum extraction in the field. Transport parameters were derived from all BTCs, based on the convection-dispersion equation and time-moment analysis.

Results and Discussion. Measured calibration relationships between ($\theta \cdot \sigma_w$) and σ_a were highly linear (Fig. 1) over the ranges in θ for all five soils used in our experiments (e.g., Fig. 2). TDR-estimated BTCs generally agreed well with those based on effluent fractions in the laboratory experiments (Fig. 3), although discrepancies were sometimes apparent during periods of low θ. Agreement between fitted transport parameters was generally good, with maximum variation in pore water velocity (v) and dispersion coefficient (D) of 11 and 17%, respectively, across all columns and measurement approaches. Field-measured BTCs using TDR at half-hour intervals were much more detailed and continuous than those based on solution sampling (Fig. 4). Measured Br and σ_w in the extracted soil solution were highly variable and scattered in comparison with TDR estimates. We noted diurnal fluctuations in TDR BTCs which apparently correspond to soil temperature. Current work includes improving the temperature compensation for specific ionic species, and evaluating the linearity of the calibration relationship over a wider range of soil water content. Measurements in KNO_3 solution at constant temperature confirm an ability to resolve changes of less than 0.001 dS m^{-1}, indicating the potential to measure meaningful changes under representative field conditions.

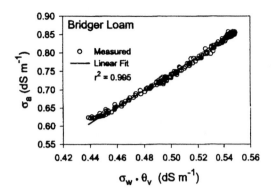

Fig. 1. Linear regression results for estimating T and σ_s during cyclic wetting and drying at constant known σ_w in repacked column.

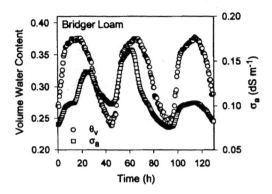

Fig. 2. Volume soil water content (θ) and bulk soil electrical conductivity (σ_a) during transport of Br pulse past TDR probe.

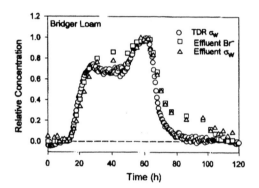

Fig. 3. Measured BTCs for repacked Bridger Loam soil column.

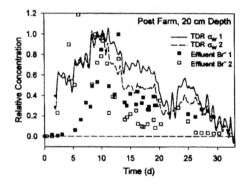

Fig. 4. Measured BTCs at 20 cm depth for Amsterdam Silt loam soil column. Data for two TDR probes and two vacuum ceramics.

References.
(1) Kachanoski, R.G., E. Pringle, and A. Ward. 1992. Field measurement of solute travel times using time domain reflectometry. Soil Sci. Soc. Am. J. 56:47-52.
(2) Vanclooster, M., D. Mallants, J. Diels, and J. Feyen. 1993. Determining local-scale solute transport parameters using time-domain reflectometry (TDR). J. Hydrol. 148:93-107.
(3) Wraith, J.M., S.D. Comfort, B.L. Woodbury, and W.P. Inskeep. 1993. A simplified waveform analysis approach for monitoring solute transport using time-domain reflectometry. Soil Sci. Soc. Am. J. 57:637-642.
(4) Risler,P.D., J.M. Wraith, and H.M. Gaber. 1995. Estimating solute transport under transient flow conditions using time domain reflectometry. Soil Sci. Soc. Am. J. (submitted).
(5) Risler,P.D., J.M. Wraith, and H.M. Gaber. 1995. Field measurement of nonsteady flow bromide transport using time domain reflectometry. Soil Sci. (submitted).
(6) Rhoades, J.D., P.A.C. Raats, and R.J. Prather. 1976. Effects of liquid-phase electrical conductivity, water content, and surface conductivity on bulk soil electrical conductivity. Soil Sci. Soc. Am. J. 40:651-655.

Stability of Wetting Fronts in Dry Homogeneous Soils
Under Low Infiltration Rates

Tzung-mow Yao, Xiaohong Du, and Jan M.H. Hendrickx[1]. *[1]Hydrology Program, Department of Earth and Environmental Science and Geophysical Research Center, New Mexico Tech, Socorro, NM 87801, USA.*

Introduction. Although our understanding of unstable wetting phenomena in soils is limited, theoretical studies supported by the results of laboratory experiments and field investigations leave no doubt that unstable wetting may play an important role during water movement in homogeneous soils and cause preferential flow paths or fingers. In three-dimensional systems with dry soils the finger diameter d of nearly saturated fingers can be calculated as:

$$d = 4.8 \frac{S_w^2}{K_s(\theta_s - \theta_0)} \frac{1}{1 - q/K_F} \qquad [1]$$

where S_w is the sorptivity of the porous medium at water entry value, K_s is the saturated hydraulic conductivity, θ_s is the saturated water content, θ_0 is the initial water content, q is the infiltration rate and 4.8 is a coefficient derived from stability analysis (Glass et al., 1991). K_F is the hydraulic conductivity inside the finger, a value assumed to be close to K_s. Eq. [1] indicates that the finger diameter will increase with increasing infiltration rates until the wetting fronts become stable at infiltration rates equal to or higher than the saturated hydraulic conductivity of the soil. Selker et al. (1992a) and Glass et al. (1989b, 1990) have presented experimental data to support Eq. [1] for infiltration rates exceeding 7 cm/h. For infiltration rates approaching zero, Eq. [1] predicts a constant small finger diameter, indicating the persistence of unstable wetting under these low rates. However, to date few experimental or theoretical studies have been carried out to experimentally verify Eq. [1] for low infiltration rates. Because low infiltration rates frequently are more representative for precipitation regimes of field soils than the high infiltration rates used in the studies reported in the literature, it is our objective to investigate the stability of wetting fronts at low infiltration rates.

Theory. In the early stages of infiltration capillary forces dominate, whereas at later times gravity becomes the dominant driving force for flow. Because unstable wetting is a gravity-driven phenomenon (Raats 1973; Parlange & Hill 1976; Philip 1975a, b), we hypothesize that no instabilities will occur during the initial stages of the infiltration process when capillary forces are dominant. For the determination of the time during which capillarity controls the infiltration process we use the gravitational characteristic time, t_{grav}, introduced by Philip (1969):

$$t_{grav} = \left(\frac{S(h)}{K_i(h) - K_o(h_o)} \right)^2 \qquad [2]$$

where S(h) is the sorptivity of the soil at supply water pressure h, $K_i(h)$ is the hydraulic conductivity at the supply soil water pressure h, and K_o is the hydraulic conductivity at the initial soil water pressure (h_o) of the soil. (4) and (3) show that at low infiltration rates t_{grav} quickly becomes much larger than the duration of the infiltration event and consequently the opportunity for gravity-driven unstable wetting fronts will disappear (3). A more rigorous approach has been followed by (1) and (2) who carried out an instability analysis of the three-dimensional Richard's equation for unsaturated water flow in porous media and derived a new expression for calculation of the finger diameter:

$$d = \pi \sqrt{\frac{D(\theta_1^*)}{B(\theta_1^*)}} \qquad [3]$$

where $D(\theta)$ is the diffusivity, $\theta_1{}^*$ is a water content close to the water content at the top boundary, and $B(\theta)$ is a function with the property $B(\theta) \rightarrow 0$ as $\theta \rightarrow \theta_o$ or $\theta \rightarrow \theta_s$, where θ_o is the initial water content and θ_s is the saturated water content. Both functions $B(\theta)$ and $\theta_1{}^*$ are determined by (1) and (2). At both high and low infiltration rates, $B(\theta_1{}^*)$ approaches to zero. Consequently, Eq. [3] predicts that d becomes very large, i.e. the wetting front becomes stable.

Materials and Methods. Lysimeter experiments were conducted in the laboratory to validate Eqs. [1] and [3]. Four different grades of sieved and air-dried perlite and quartz sand were used as the experimental material. Water was applied by a sprinkler system at rates within the range of natural precipitation rates in New Mexico. Experiments were conducted in small lysimeters (diameter 30 cm, height 50 cm) as well as a large one (diameter 100 cm, height 150 cm). For more details on our method we refer to Yao and Hendrickx (1995).

Results and Discussion. Our experimental results in the 14-20 coarse sand show that, for infiltration rates varying between 0.3 and 12 cm/h, finger diameters remain more or less constant. This observation agrees with Eq. [1]. However, at infiltration rates lower than 0.12 cm/h, the coarse sand experiments show that the wetting fronts became stable. For rates between 0.3 and 0.12 cm/h, the wetting is semi-stable; that is, there is incomplete wetting without distinct development of fingers (Fig. 1). A similar trend is observed in the experimental results of sands with grain sizes of, respectively, 0.841-0.594 and 0.594-0.42 mm. This phenomenon has not been observed in previous experimental studies and is not predicted by Eq. [1]. However, our Eq. [3] correctly predicts finger diameters under high, intermediate, and low infiltration rates (Fig. 1). The experimental and theoretical results of this study under infiltration rates similar to natural precipitation intensities may explain why unstable wetting has not been observed more frequently in wettable field soils (Hendrickx and Yao 1995).

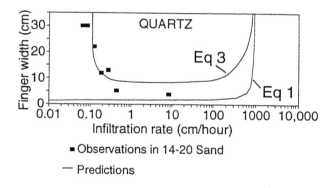

Fig. 1. The relationship between finger diameter and infiltration rate. Experimental results in 14-20 (■) sand and finger diameters calculated with Eqs. [1] and [3] for 14-20 sand.

Literature Cited.
We only present references of our papers in press or submitted. All other references can be found in the forthcoming paper by Yao and Hendrickx (1995)

(1) Du, X. and J.M.H. Hendrickx. 1995a. A method to solve the one-dimensional vertical flow Richard's equation. Submitted to Water Resources Research.
(2) Du, X. and J.M.H. Hendrickx. 1995b. Instability analysis of the three-dimensional Richard's equation for unsaturated water flow in porous media. To be submitted to Water Resources Research.
(3) Hendrickx, J.M.H. and T. Yao. 1995. Prediction of wetting front stability in dry homogeneous field soils under non-ponding infiltration. Geoderma, in press.
(4) Yao, T. and J.M.H. Hendrickx. 1995. Stability of wetting fronts in homogeneous soils under low infiltration rates. Soil Science Society of America Journal, in press.

Comparison of Aggregate Breakdown in Pasture Topsoils

E. Zanini[1], E. Bonifacio [1] and D.R. Nielsen [2]. [1] *Dipartimento di Valorizzazione e Protezione delle Risorse Agroforestali, Università di Torino, I-10100 Torino, Italy.* [2] *Dept. of Land, Air and Water Resources, University of California, Davis, CA 95616, USA.*

Introduction. Water erosion losses depend on soil structural stability (1,2). The aggregate stability of the upper horizons is well related to land use and cropping history - cultivated versus native species or pasture (3,4). Because most methods of measuring soil aggregation are based upon the assumption that soils possess a minimum stability under water-saturated conditions, it is common to measure the resistance of the soil aggregates to breakdown when they are subjected to mechanical abrasion in water (e.g. 5,6). The objective of this work was to represent aggregate stability during wet sieving with a simple mathematical model in order to compare different behaviors under similar management of soils.

Material and Methods. Twenty-four topsoils (A or AC horizon) from highly erodible pasture lands in Southern Apennines (Camastra valley, Italy, 40°22'- 40°35' N and 15°42'-16°01' E) were analyzed. All of the pastures had the same cropping history and similar topographic conditions (slope lenght of about 500 m and steepness ranging between 28 and 35 %) . The soils, *Lithic and Typic Ustorthents*, were poorly developed on six different parent materials belonging to a Tertiary complex (dolomitic limestones, shales, conglomerates, flysch, marls and calcareous sandstones).
After air-drying and low-vacuum rewetting affregates of 1-2 mm diameter (7), their stability was determined by wet sieving (8). Stable aggregates (>0.25 mm diameter) resisting abrasion after sieving times of 5, 10, 15, 20, 40 and 60 min were ascertained by correcting for coarse sand.
Inasmuch as the wet sieving data showed an exponential trend, we assumed that the breakdown of the aggregates followed the first order decay model:

$$y(t) = a+b \, [1-\exp(-t/c)] \tag{1}$$

where **y** is the aggregate loss; **t** the time of wet sieving (abrasion); **a** the nugget-like value consistent with the "explosion" of the pre-wetted aggregates when effectively saturated in water; **b** the abrasion loss of aggregates and **c** the parameter that accounts for the rate of the abrasion loss. Note that **(a+b)** is the maximum loss of aggregates (explosion plus abrasion). The range of the asymptote to **(a+b)** is arbitrarily defined to be **3c** corresponding approximately to the time at which **0.95(a+b)** is reached (9). We estimated the values of **a**, **b** and **c** using an iterative non linear regression procedure until the residual sum of squares difference between iterations was $<10^{-8}$. The rate of aggregate breakdown is described by the derivative:

$$y'(t) = (b/c) \, \exp(-t/c) \tag{2}$$

and $y'(t_o)$ is the initial rate of the abrasion loss that is indipendent from the "explosion ". The "half life" of the aggregate loss is simply

$$t_{50} = -c \; \lg[1-(y_{50} - a)/b] \tag{3}$$

Results and Discussion. The wet sieving abrasion data for each of all sampling locations were adequately described by the regression model with values of r^2 always being greater than 0.95. Significant differences for the parameters were found from a one-way ANOVA by lithology (Table1). Figure 1 shows the averaged curves for each of the lithologies. Soil aggregates on conglomerates were the weakest while those on limestones and shales were the most "explosive".

Conclusions. With the regression model successfully describing soil aggregate breakdown in more than 200 samples analyzed, we believe that it can be used as a practical tool to ascertain potential water erosion losses from different soils and those from the same soil for different land use and cropping strategies. Here, for different *Lithic and Typic Ustorthents* under pasture and having the same cropping history and similar topographic conditions, we have learned that the dynamics of aggregate breakdown is related to lithology. Presently, we are investigating how well the model can discriminate aggregate breakdown of different soils in the same cropping system and that of the same soil under different cropping systems.

Table 1: Model parameters averaged by lithology and ANOVA

Lithology	a g/100g	a+b g/100g	3c min	$\dot{y}(x_0)$ g/min	x_{50} min
Limestones	1.6 a	47.7 c	53 a	3 c	11 a
shales	1.6 a	38.0 c	34 b	4 c	7 b
marls	0.4 b	17.7 d	46 a	1 c	10 a
conglomerates	0.0 b	92.6 a	6 c	46 a	1 b
flysch	0.2 b	72.3 b	16 b	16 b	4 b
sandstones	0.8 b	63.8 b	26 b	14 b	6 b

Different letters denote significant differences among groups.

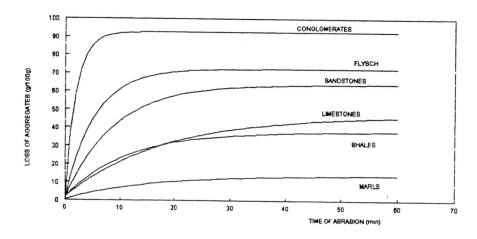

Figure 1: Decay of the aggregates according to the lithology

Literature Cited.

(1) Anderson, H.W., 1951, Physical characteristics of soil related to erosion. .Soil Water Conserv., 6: 129-133.
(2) Stern, R., Eisenberg, B.E. and Laker, M.C., 1991, Correlation between microaggregate stability and soil surface susceptibility to runoff and erosion. S.Afr.Tydskr.Plant Grond, 8(3), 136-140.
(3) Sullivan, L.A., 1990, Soil organic matter, air encapsulation and water-stable aggregation. J.Soil Sci., 41, 529-534.
(4) Haynes, R.J., 1993, Effect of sample pretreatment on aggegate stability measured by wet sieving or turbidimetry on soils of different cropping history. J.Soil Science, 44, 261-270.
(5) Yoder, R.E., 1936, A direct method of aggregate analysis and a study of the physical nature of soil erosion losses. Agronomy J., 28:237-351.
(6) Beare, M.H. and Bruce R.R., 1993, A comparison of methods for measuring water-stable aggregates: implication for determining environmental effects on soil structure. Geoderma, 56, 87-104.
(7) Dickson, J.B., Rasiah, V., Groenevelt, P.H. 1991, Comparison of four prewetting techniques in wet aggregate stability determination. Can.J.Soil.Sci., 71(1):67-72.
(8) Kemper, W.D., Resenau, R.C., 1986, Aggragate stability and size distribution. In: A. Klute (ed), Methods of soil analysis. Part I. Physical and mineralogical methods. Am.Soc. of Agron./Soil Sci.Soc of A., Madison, WI, 425442.
(9) Isaaks, E.H. and Srivastava, R.M., 1989, An Introduction to Applied Geostatistics. Oxford University Press, New York, pp.561.

This work was granted by the National Research Council, Rome, Italy (C.N.R.-R.A.I.S.A. subproject 1).

Alphabetical Index of Oral and Poster Presenters

Presenter's Name	Oral Presentation or Poster Number(s)	Presenter's Name	Oral Presentation or Poster Number(s)
L. M. Abriola	oral presentation	D. L. Freyberg	25
J. D. Albertson	1, 7, 52	W. H. Frohlich	21
G. B. Allison	oral presentation	T. Frueh	3
G. Alvenas	2	H. M. Gaber	22, 77
M. C. Amacher	58	W. R. Gardner	oral presentation
M. A. Anderson	31	J. Garison	59
L. Andreu	3	G. W. Gee	oral presentation
M. D. Ankeny	4	P. Germann	23
L. R. Angelocci	55	R. W. Gillham	63
O. O. S. Bacchi	51, 55	C. Gonzales	74
A. C. Bagtzoglou	67	S. A. Grant	24
D. A. Barry	14, 66, oral presentation	W. G. Gray	oral presentation
P. Boivin	11	T. R. Green	25
W. J. Bond	75	M. E. Grismer	9
E. Bonifacio	1, 79	P. H. Groenevelt	26
P. Bottomley	59	W. Hamminga	30
J. Bouma	oral presentation	M. R. Hara	27
S. A. Bradford	38	P. Hairsine	32
E. Braudeau	5, 6, 11	R. F. Harris	oral presentation
K. L. Bristow	75	Th. Harter	28
J. J. B. Bronswijk	30	R. Haverkamp	71, oral presentation
M. L. Brusseau	oral presentation	M. Hayashi	29
M. E. Burkhard	4	D. A. Heeraman	10
A. T. Cahill	7, 52	A. W. J. Heijs	56
S. F. Carle	8	J. M. H. Hendrickx	61, 78
H. Castellaw	67	R. F. A. Hendriks	30
M. A. Celia	oral presentation	Z. R. Hinedi	31
A. C. Chang	31	W. L. Hogarth	32
J. Chen	9	J. W. Hopmans	3, 9, 10, 17
V. Clausnitzer	10	J. R. Hunt	33
J. E. Constantz	25	M. P. Inskeep	22
Y. Coquet	11	I. K. Iskandar	58
H. P. Cresswell	75	S. Ita	46
S. Crestana	12	T. J. Jackson	52, oral presentation
R. H. Cuenca	13	P.-E. Jansson	39
P. J. Culligan-Hensley	14, oral presentation	C. D. Johnston	66
B. S. Das	15	M. Joschka	oral presentation
J. H. Dane	41	W. A. Jury	oral presentation
L. W. Dekker	16, 48, 56	Z. J. Kabala	oral presentation
D. Dourado-Neto	17	R. G. Kachanoski	20
X. Du	78	C. S. Kao	33
D. E. Elrick	18, 26	F. Kastanek	34
E. T. Engman	oral presentation	B. A. Keating	75
S. Essert	3	S. F. Kelly	13
Y. El-Farhan	oral presentation	J. S. Kern	35
B. Faybishenko	19	D. Kirkham	36
P. Ferre	20	M. B. Kirkham	36
J. Feyen	42, 74, 76	G. J. Kluitengerg	15
S. Finsterle	19	M. Kutilek	37
G. E. Fogg	8, 21	H. W. Langner	22

Presenter's Name	Oral Presentation or Poster Number(s)	Presenter's Name	Oral Presentation or Poster Number(s)
F. J. Leij	38	D. L. Rudolph	20, 29
E. Lewan	39	G. Sander	32
H. S. Lin	40	T. Sawyer	59
I. Lisle	32	B. R. Scanlon	oral presentation
H. H. Liu	41	L. J. Schwankl	3
Y. Liu (UCD)	9	H. D. Scott	40
Y. Liu (UCB)	oral presentation	K. M. Scow	oral presentation
J. MacIntyre	3	H. M. Selim	58
L. Ma	58	J. S. Selker	59
R. S. Maier	43	M. Seyfried	60
D. Mallants	42, 74	K. R. Sheets	61
D. Marks	35	M. I Sheppard	18
M. Mata	52	A. M. Shurbaji	62
A. S. Mayer	45	J. E. Smith	63
J. M. McKimmey	40	W. E. Soll	64
D. McLaughlin	oral presentation	G. Sposito	oral presentation
P. Meyer	47	D. E. Stangel	13
D. Misra	43, 44, 48, 49	J. L. Starr	65
R. J. Mitchell	45	T. S. Steenhuis	16, oral presentation
C. Montemagno	46	D. A. Steffy	66
B. Moore	52	J. S. Steude	10
H. J. Morel-Seytoux	47	S. A. Stothoff	67
J. Mount	21	E. A. Sudicky	oral presentation
D. J. Mulla	oral presentation	J. Szilagyi	68
A. Nadler	18	K. C. Tarboton	69
N. V. Nguyen	44, 48, 49	A. Thomsen	70
J. L. Nieber	43, 44, 48, 49, 73	D. J. Timlin	65
D. R. Nielsen	1, 17, 79	N. Toride	42
P. Nkedi-Kizza	50	J. Touma	6, 11, 71
K. Novins	35	A. Tuli	3
V. Nzengung	50	H. J. Turin	72
B. P. Odell	26	S. W. Tyler	oral presentation
J. C. M. de Oliveira	51	M. Vanclooster	74
C. Oman	57	G. van den Eertwegh	73
P. O'Neill	52	E. van den Elsen	56
K. Oostindie	30	J. Vanderborght	74
I. C. Paltineanu	65	G. van der Kamp	29
L. Pan	oral presentation	M. Th. van Genuchten	42, 74, oral presentation
J.-Y. Parlange	14, 16, 32, oral	C. M. P. Vaz	51
M. B. Parlange	7, 17, 52, 68	K. Verburg	75
K. D. Penell	oral presentation	E. L. Vold	72
G. Persson	53	J. E. Waatsonn	oral presentation
J. E. Pilotto	55	R. J. Wagenet	oral presentation
A. N. D. Posadas	12	W. W. Wallender	54, 69
M. A. Prieksat	4	Z. Wang	76
M. E. Probert	75	A. W. Warrick	oral presentation
D. Purkey	54	W. J. Weber, Jr.	oral presentation
M. Quintard	oral presentation	O. Wendroth	oral presentation
K. Reichardt	51, 55	S. Whitaker	oral presentation
P. D. Risler	77	P. J. Wierenga	oral presentation
C. J. Ritsema	16, 48, 56	J. M. Wraith	22, 77
D. B. Rogers	72	T. Yao	78
D. E. Rolston	oral presentation	E. Zanini	1, 79
C. W. Rose	32	D. Zhang	28
H. Rosqvist	57	H. Zhu	58